Our Global Future

Revised First Edition

Edited by David Larom, Ph.D

San Diego State University

San Diego, CA

Bassim Hamadeh, CEO and Publisher
Christopher Foster, General Vice President
Michael Simpson, Vice President of Acquisitions
Jessica Knott, Managing Editor
Kevin Fahey, Cognella Marketing Manager
Jess Busch, Senior Graphic Designer
Seidy Cruz, Acquisitions Editor
Sarah Wheeler, Project Editor
Stephanie Sandler, Licensing Associate

First published in the United States of America in 2013 by Cognella, Inc.

Printed in the United States of America

ISBN: 978-1-62131-461-5 (pbk)

www.cognella.com   800.200.3908

Contents

Introduction

I have tried to make this book a little more interesting, more personal, and in some cases more upbeat than most descriptions of our global sustainability crisis. It's easy to emphasize the troubling facts, to the point where depression and paralysis set in. You've heard it before: "Global warming is frying the planet. The Pacific Ocean is a giant garbage patch. All the coral is dying, the food is contaminated with bacteria and there's not enough of it, there are too many people, China is choking the world with coal emissions, blah, blah, blah. …" Well, not quite blah, blah, blah since there is some truth to all these statements. Nonetheless, I would not be writing this intro to, nor editing the contents of, a book called *Our Global Future* unless I thought we had one! Don't get me wrong; we'll be looking at a lot of tough problems—but the fun thing about problems is solving them. Human ingenuity is boundless, and I believe our future is bright if we can engage creatively with pollution, overpopulation, famine, erosion, biodiversity loss, and all the other troubling consequences of *Homo sapiens'* rise to prominence over the last few thousand years.

That said, you won't find this book a cheery advocate for technology as the solution to all our problems. Most of our pressing problems are the result of a system of linearized thinking, expressed as industrial production wedded to a growth-oriented capitalist economy. Nuclear power and genetic engineering are part of this worldview. They are not, in my view, going to save the planet; at best they will be expensive sidelines to the real game, that of moving from growth to maintenance as we mature as a species. A child should grow; an adult should stay the same size. At some point the global economy has to *stop* growing. How we reach that point is the entire game. A body of research suggests we may even have overshot the planetary carrying capacity, meaning we are simply consuming more stuff and dumping more garbage than the planet can handle. When I was a college freshman, I gained ten pounds on the free dorm food; when I was a grad student and got my first credit card I borrowed too much money and had to tighten my belt and pay it off. The human race is like a slightly spoiled teenager who needs to rein herself in. We need more care and self-control. It will be painful to admit we can't have certain things, but the feeling of living within one's means—which is the only meaning of "sustainability" that matters—can be a joyous affirmation of values that have kept communities alive for generations.

Actually, technology *is* a big part of the solution, but it is *appropriate* technology; tech that is sized for the job at hand and often quite small and specialized. Bigger isn't always better. Small farmers take better care of their five to ten acres, and produce more and better food per unit input than corporations running multi-thousand-acre, industrial-agricultural food factories (the UN's IAASTD study[1] backs me up on this point). Countries that depend on other countries thousands of miles away for their energy supplies must do *anything* to keep the energy coming—including install puppet dictators, support tyrants, and go to war to maintain supply lines. Globe-spanning "just-in-time delivery" of industrial products, food, and energy may be "efficient" from a production standpoint, but it is highly susceptible to natural disasters and terrorist attacks. NB: The same can be said of the giant system of dams and water supply lines that spans California and the American West. We in Southern California are one giant earthquake away from dying of thirst!

One theme that emerges from this book, therefore, is that the new technologies that save us may actually be ancient, small scale, and based on common sense. Catch the rainwater on your roof; generate and distribute power from many small sources over a robust network; know your farmer,

not to mention your neighbor. Share knowledge and wealth. With the exception of the energy network, these principles have been the foundation of every successful tribe and village since humanity's beginning. They have also been forgotten in the current rush to "globalize," and great waste was the result. The 1980s gave us an enjoyable movie called *Back to the Future*; this collection suggests that "Forward to the Past" might be a healing direction for humanity and our beloved planet.

The so-called developing world is another major theme of this book. Most readers and writers of books on sustainability are from the more affluent countries of the "developed world"; there has therefore been an unconscious bias toward discussing "our" issues as opposed to "theirs." The rise of China is changing this, but even then, many of the stories we read in the press concern "them" becoming "us," so to speak. It will be difficult—nay, impossible—for the planet to support a billion-plus Chinese living affluent U.S. lifestyles. The headline story of the 21st century is the industrialization and urbanization of the developing world, but there's another story, at least as important, about the people who are being abandoned along with their traditions in the headlong rush to wasteful and polluting modernization. The industrial paradigm is squeezing Asian, African, and Latin American peasant farmers and fishermen out of existence, much as it did European peasants during the Industrial Revolution.

There are two main reasons to pay particular attention to these folks. The first is simple compassion: *They suffer the most.* They are the most vulnerable, their lifestyle is the least respected, and they are the most abused of the world's major population groups (the hunter-gatherers, tragically, have essentially been wiped out). The great educator Paolo Freire emphasized the "option for the poor"; the American Disabilities Act has ensured that we allow our most challenged citizens to access the same goods, services, and facilities that able-bodied

---

1 United Nations, 2008, available at http://www.agassessment.org/. Printed as McIntyre, B.D., Herren, H.R., Wakhungu, J. & Watson, R.T. eds (2009) "Agriculture at a Crossroads: International Assessment of Agricultural Knowledge, Science and Technology for Development—Global Report." Island Press, Washington, DC.

adults can easily enjoy. Simple humanity requires that we similarly assist the peasant.

The second reason will appeal even to the most Machiavellian among us: *These peasants feed the world sustainably.* Traditional farmers and fishermen are better resource stewards, since their communities have depended on them for centuries or millennia to sustain them. As humanity rushes into the cities, the countryside is depopulated. All over the developing world, corporate "farmers" are rushing to fill this vacuum, with profoundly destabilizing effects. The corporate farm and the trawler, in contrast to the village farmer and fisherman, must focus on the bottom line, on quarterly profit, or they will be outcompeted. The land and the ocean must produce the maximum food in the minimum time, leading to terrible resource drawdowns.

At an environmental conference some 20 years ago, I had an edifying discussion with one of the presenters. He had been talking about how whaling must be stopped entirely to prevent the hunted species from being driven to extinction. "Now listen," I said to him. "I agree with you that whaling is bad. Personally, I think whales are way too intelligent to kill for food. But for the sake of argument, let's just put that aside. Suppose you could kill 10% of the whales every year, but they reproduced fast enough that their numbers remained the same. Wouldn't whaling then be sustainable?"

"Your problem," said my acquaintance, "is that you are thinking like a whaler and not like a capitalist. As a whaler, you kill whales for a living.

You want to keep doing this and to pass this lifestyle on to your kids. You will take your 10% permanent harvest rather than run out of whales. A capitalist has a different set of calculations. If a capitalist can instead make 20% or 30% by killing every whale on the planet in five or ten years, he'll do it!"

"Nonsense!" I replied. "He'd put himself out of business. Only an idiot would destroy his own livelihood."

"You just don't get it!" the presenter shot back. "*He would have maximized his profit.* He's not a whaler. When the world runs out of whales, he'll just reinvest in something else."

That shut me up. Such is the logic, if you can call it that, of our modern economic system. People often speak of sustainability as if it were a lifestyle choice, a little set of "feel-good" activities consisting of recycling, driving a Prius, and buying organic food. Don't get me wrong; I do all of these. But humanity is not going to be rescued from collapse until there is a revolution in our global economic system—or, as the Green Party used to put it, a *half*-revolution, since if you do a *full* revolution, you're still going in the same direction. It's likely none of us know what such a (half) revolution will look like until it occurs. I am skeptical it will merely be a greener capitalism. History has probably established it will not look like so-called communism, which is mere state capitalism and even more damaging to the global commons. Whatever it is, if you're under 30, chances are good it will happen in your lifetime. *You will help make it happen.* Enjoy!

A Short History of Progress

By Ronald Wright

Introduction

We begin this book with "The Great Experiment," an early chapter from Ronald Wright's *A Short History of Progress*, about the transition from hunting and gathering to farming—the so-called Neolithic Revolution. In a course called "Our Global Future," why go so far into the past? The answer is simple: *Because history repeats itself* (or if not, at least it rhymes, as Twain once quipped). About 10,000 years ago, hunter-gatherers got too good at hunting and gathering, and their population grew beyond the ability of the land to support them with meat and wild forage. Hunters put themselves out of the hunting business and had to learn to farm. As farming improved, the human population grew far more rapidly, and another food/population crisis was reached a few centuries ago. Since then, the Industrial Revolution has supplanted traditional, organic farming with highly mechanized monoculture farming dependent on fossil fuel inputs. We now have seven billion people on the Earth, and the population is forecast to grow to nine billion by 2050. The last few years have seen a tripling of global grain prices and the return of the spectre of mass famine to the world.

For thousands of years now, the dominant human paradigm has been one of growth. The global economy is entirely dependent on continued growth; every day we hear worrying news reports questioning: Has the economy grown this quarter? Is the stock market up or down today? For the ancient hunters, growth ended when the big game was killed off. For traditional farmers, growth ended when the good land was all taken. For industrial civilization, growth is ending. For the first time, humans are using a significant portion of the productive capacity of the entire planet to feed themselves. It is, therefore, highly unlikely we will produce ourselves out of this third global food crisis. The only rational response is to begin to limit ourselves. We will need to learn some form of moderation.

That's the bad news. The good news is that, throughout history, pockets of humanity have encountered local resource shortages and learned the discipline of limitation. We will see numerous such examples as this book progresses. To give a personal example: I was born in 1961, during the heyday of limitlessness, yet my parents were children of the Great Depression. They, and my grandparents and other older relatives, learned painful lessons about conservation just to survive. To give but one example, my great-aunt Melita saved left-over slivers of bath soap to wash the dishes (I won't scare you with stories of having to eat organ meats—think *menudo*—instead of flesh; just remember these are incredibly nutritious!). Over the past decades, we have had numerous victories over the perennial philosophy of waste. Cars pollute a lot less now; no U.S. rivers have caught fire as the Cuyahoga did in 1969. The Clean Air and Clean Water Acts, signed by the Republican president Nixon, have saved wildlife, restored some of the natural beauty of our nation, and slowed the mass poisoning of our citizens by industrial toxins.

Conservation and care are, in short, a more sustainable life philosophy than limitless growth. "Growth for growth's sake is the ideology of the cancer cell," as Edward Abbey famously quipped.

But I digress (a worthy digression, though, since it sets the mood and helps you understand why you are reading this chapter). To return to Ronald Wright: *A Short History of Progress* is an elegiac, beautifully written book, a paean to what we have lost. The growth paradigm has not just been unsustainable; it has also been undemocratic, because political structures have changed radically since hunter-gatherer times. A tribe of 30 or so relatives has to cooperate, share, and agree on things. Karl Marx called this arrangement "primitive communism," but lest anyone paint this commentary with a red brush, let me also point out that others have lived by this creed: "All the believers were one in heart and mind. No one claimed that any of their possessions was their own, but they shared everything they had." (*Bible*, New International Version, Acts 4:32). The point is this: Small bands of like-minded humans share, but Wright shows that as "progress" occurs and group sizes grow, "civilization" can be downright uncivilized to its citizens. The parallel historical occurrence of unsustainability and oppression is a recurring theme that rings as true today as it did 6000 years ago in Sumeria.

The Great Experiment

To draw a rough analogy between two unconnected eras of very different length and complexity, there are certain resemblances between this end-time of the Old Stone Age and the past half millennium of Western "discovery" and conquest. Since A.D. 1492, one kind of civilization—the European—has largely destroyed and displaced all others, fattening and remaking itself into an industrial force in the process (a point I shall return to in a later chapter). During the Upper Palaeolithic, one kind of human—the Cro-Magnon, or *Homo sapiens*—multiplied and fanned out around the world, killing, displacing, or absorbing all other variants of man, then entering new worlds that had never felt a human foot.

By 15,000 years ago at the very latest—long before the ice withdraws—humankind is established on every continent except Antarctica. Like the worldwide expansion of Europe, this prehistoric wave of discovery and migration had profound ecological consequences. Soon after man shows up in new lands, the big game starts to go missing. Mammoths and woolly rhinos retreat north, then vanish from Europe and Asia. A giant wombat, other marsupials, and a tortoise as big as a Volkswagen disappear from Australia. Camels, mammoth, giant bison, giant sloth, and the horse die out across the Americas. A bad smell of extinction follows *Homo sapiens* around the world.

Not all experts agree that our ancestors were solely to blame. Our defenders point out that we hunted in Africa, Asia, and Europe for a million years or more without killing everything off; that many of these extinctions coincide with climatic upheavals; that the end of the Ice Age may have come so swiftly that big animals couldn't adapt or migrate. These are good objections, and it would be unwise to rule them out entirely. Yet the evidence against our ancestors is, I think, overwhelming. Undoubtedly, animals were stressed by the melting of the ice, but they had made it through many similar warmings before. It is also true that earlier people—*Homo erectus,* Neanderthals, and early *Homo sapiens*—had hunted big game without hunting it out. But Upper Palaeolithic people were far better equipped and more numerous than their forerunners, and they killed on a much grander scale. Some of their slaughter sites were almost industrial in size: a thousand mammoths at one; more than 100,000 horses at another. "The Neanderthals were surely able and valiant in the chase," wrote the anthropologist William Howells in 1960, "but they left no such massive bone yards as this." And the ecological moral is underlined more recently by Ian Tattersall. "Like us," he says, "the Cro-Magnons must have had a darker side."

In steep terrain, these relentless hunters drove entire herds over cliffs, leaving piles of animals to rot, a practice that continued into historic times at places such as Head-Smashed-In Buffalo Jump, Alberta. Luckily for bison, cliffs are rare on the great plains. But there would be no limit to the white man's guns that reduced both buffalo and Indian to near extinction in a few decades of the nineteenth century. "The humped herds of buffalo," wrote

Herman Melville, "not forty years ago, overspread by tens of thousands the prairies of Illinois and Missouri ... where now the polite broker sells you land at a dollar an inch." Land at a dollar an inch: now *that* is civilization.

Modern hunter–gatherers—Amazonians, Australian Aboriginals, Inuit, Kalahari "bushmen"—are wise stewards of their ecologies, limiting their own numbers, treading lightly on the land. It is often assumed that ancient hunters would have been equally wise. But archaeological evidence does not support this view. Palaeolithic hunting was the mainstream livelihood, done in the richest environments on a seemingly boundless earth. Done, we have to infer from the profligate remains, with the stock-trader's optimism that there would always be another big killing just over the next hill. In the last and best-documented mass extinctions—the loss of flightless birds and other animals from New Zealand and Madagascar—there is no room for doubt that people were to blame. The Australian biologist Tim Flannery has called human beings the "future-eaters." Each extermination is a death of possibility.

So among the things we need to know about ourselves is that the Upper Palaeolithic period, which may well have begun in genocide, ended with an all-you-can-kill wildlife barbecue. The *perfection* of hunting spelled the *end* of hunting as a way of life. Easy meat meant more babies. More babies meant more hunters. More hunters, sooner or later, meant less game. Most of the great human migrations across the world at this time must have been driven by want, as we bankrupted the land with our moveable feasts.

The archaeology of western Europe during the final millennia of the Palaeolithic shows the grand lifestyle of the Cro-Magnons falling away. Their cave painting falters and stops. Sculptures and carvings become rare. The flint blades grow smaller, and smaller. Instead of killing mammoth they are shooting rabbits.

In a 1930s essay called "In Praise of Clumsy People," the waggish Czech writer Karel Capek observed: "Man ceased to be a mere hunter when individuals were born who were very bad hunters." As someone once said of Wagner's music, Capek's remark is better than it sounds. The hunters at the end of the Old Stone Age were certainly not clumsy, but they were bad because they broke rule one for any prudent parasite: *Don't kill off your host.* As they drove species after species to extinction, they walked into the first progress trap.

Some of their descendants—the hunter–gatherer societies that have survived into recent times—would learn in the school of hard knocks to restrain themselves. But the rest of us found a new way to raise the stakes: that great change known to hindsight as the Farming or Neolithic "Revolution."

Among hunters there had always been a large number of non-hunters: the gatherers—mainly women and children, we suppose, responsible for the wild fruits and vegetables in the diet of a well-run cave. Their contribution to the food supply became more and more important as the game died out.

The people of that short, sharp period known as the Mesolithic, or Middle Stone Age, tried everything: living in estuaries and bogs; beachcombing; grubbing up roots; and reaping wild grasses for the tiny seeds, a practice with enormous implications. So rich were some of these grasses, and so labour-intensive their exploitation, that settled villages appear in key areas *before* farming. Gatherers began to notice that seeds accidentally scattered or passed in droppings would spring up the following year. They began to influence the outcome by tending and enlarging wild stands, sowing the most easily reaped and plumpest seeds.

Such experiments would eventually lead to full agriculture and almost total dependence on a few monotonous staples, but that was several thousand years away; at this early time, the plant-tenders were still mainly gatherers, exploiting a great variety of flora, as well as any wild game and fish they could find. At Monte Verde in Chile, for example, a permanent village of rectangular wooden huts was in place by 13,000 years ago, sustained by hunting camelids, small game, and soon-to-be-extinct mastodon; but the remains include many wild vegetables, not least potato peelings. Although Monte Verde is one of the earliest human sites anywhere in the Americas, it shows a mature and intimate knowledge of local plants, several of which would eventually become the founding crops of Andean civilization.

Like the accumulation of small changes that separated us from the other great apes, the Farming Revolution was an unconscious experiment, too gradual for its initiators to be aware of it, let alone to foresee where it would lead. But compared with all earlier developments, it happened at breakneck speed.

Highly important, for what it tells us about ourselves, is that there was not one revolution but many. On every continent except Australia, farming experiments began soon after the regime of the ice released its grip. Older books (and some recent ones) emphasize the importance of the Middle East, or the Fertile Crescent, which in those days stretched from the Mediterranean shore to the Anatolian plateau and the alluvial plains of Iraq. All the bread-based civilizations derive their staples from this area, which gave us wheat, barley, sheep, and goats.

It is now clear that the Middle East was only one of at least four major regions of the world where agriculture developed independently at about the same time. The others are the Far East, where rice and millet became the main staples; Mesoamerica (Mexico and neighbouring parts of Central America), whose civilizations were based on maize, beans, squash, amaranth, and tomatoes; and the Andean region of South America, which developed many kinds of potato, other tubers, squash, cotton, peanuts, and high-protein grains such as quinoa. In all these heartlands, crop domestication appears between 8,000 and 10,000 years ago. Besides these Big Four, there are about a dozen lesser founding areas around the world, including tropical Southeast Asia, Ethiopia, the Amazon, and eastern North America, which gave us, respectively, the banana, coffee, manioc, and the sunflower. Unconnected peoples sometimes developed the same plants: cotton and peanuts are each of two kinds, developed simultaneously in the New World and the Old.

Animal domestication is harder to document, but at about the same time people were developing crops, they learned that certain herbivores and birds could be followed, corralled, and killed at a sustainable rate. Over generations these animals grew tame enough, and dim-witted enough, not to mind the two-legged serial killer who followed them around. Hunting became herding, just as gathering grew into gardening.

Sheep and goats were the first true domesticates in the Middle East, starting about 8000 B.C. Domestic camelids—early forms of the llama and alpaca, used for pack trains and wool, as well as for meat—appear in Peru during the sixth millennium B.C., about the same time as cattle in Eurasia, though neither camelids nor early cattle were milked. Donkeys and horses were tamed by about 4000 B.C. Craftier creatures such as dogs, pigs, and cats had long been willing to hang around human settlements in return for scraps, slops, and the mouse boom spurred by granaries. Dogs, which may have been tamed for hunting back in the Palaeolithic, are found with human groups throughout the world. In cold weather, they were sometimes used as bedwarmers. In places such as Korea and Mexico, special breeds were kept for meat. The chicken began its sad march towards the maw of Colonel Sanders as a gorgeously feathered Asian jungle fowl, while Mexico domesticated the turkey. Along with the llama and alpaca, Peruvians kept muscovy ducks and the lowly but prolific guinea pig—which even made a cameo appearance on the menu of Christ's Last Supper in a colonial painting.

As the eating of guinea pigs and chihuahuas suggests, the Americas were less well-endowed with domesticable animals than the Old World. But the New World compensated by developing a wider and more productive range of plants. Peru alone had nearly forty major species. Such plants eventually supported huge native cities in the Americas, and several of them would transform the Old World's nutrition and economics when they were introduced there—a matter I shall discuss in the final chapter.

The more predictable the food supply, the bigger the population. Unlike mobile foragers, sedentary people had little reason to limit the number of children, who were useful for field and household tasks. The reproductive rate of women tended to rise, owing to higher levels of body fat and earlier weaning with animal milk and cereal baby food. Farmers soon outnumbered hunter–gatherers—absorbing, killing, or driving them into the surrounding "wilderness."

At the beginning of the Upper Palaeolithic, when our modern subspecies emerged by fair means or foul as the earth's inheritors, we numbered perhaps a third of a million all told. By 10,000 years ago, on the eve of agriculture and after settling all habitable continents, we had increased to about 3 million; and by 5,000 years ago, when farming was established in all the founding regions and full civilization had begun in Sumer and Egypt, we may have reached between 15 and 20 million worldwide.

Such figures are merely educated guesswork, and everything else I have just said is, of course, an oversimplification. The change to full-time farming took millennia, and early results were not always promising, even in a core zone such as the Middle East. Neolithic Jericho was tiny, a mere four acres in 8000 B.C., and it took another 1,500 years to reach ten acres. The Turkish site of Catal Huyuk, the largest settlement in the Fertile Crescent between 7000 and 5500 B.C., covered only one twentieth of a square mile (or thirty-two acres), and its inhabitants depended on wild game for much of their protein. As any rural Canadian knows, hunting continues among farmers wherever it's fun or worthwhile, and this was especially true in the Americas and parts of Asia where domestic animals were scarce. Nevertheless, the pace of growth accelerated. By about 5,000 years ago, the majority of human beings had made the transition from wild food to tame.

In the magnitude of its consequences, no other invention rivals farming (except, since 1940, the invention of weapons that can kill us all). The human career divides in two: everything before the Neolithic Revolution and everything after it. Although the three Stone Ages—Old, Middle, and New—may seem to belong in a set, they do not. The New Stone Age has much more in common with later ages than with the millions of years of stone toolery that went before it. The Farming Revolution produced an entirely new mode of subsistence, which remains the basis of the world economy to this day. The food technology of the late Stone Age is the one technology we can't live without. The crops of about a dozen ancient peoples feed the 6 billion on earth today. Despite more than two centuries of scientific crop-breeding, the so-called green revolution of the 1960s, and the genetic engineering of the 1990s, not one new staple has been added to our repertoire of crops since prehistoric times.

Although the New Stone Age eventually gave rise to metalworking in several parts of the world, and to the Industrial Revolution in Europe, these were elaborations on the same theme, not a fundamental shift in subsistence. A Neolithic village was much like a Bronze or Iron Age village—or a modern Third World village, for that matter.

The Victorian archaeological scheme of classifying stages of human development by tool materials becomes unhelpful from the Neolithic onward. It may have some merit in Europe, where technology was often linked to social change, but is little help for understanding what happened in places where a lack of the things our technocentric culture regards as basic—metal, ploughs, wheels, etc.—was ingeniously circumvented, or where, conversely, their presence was inconsequential. For example, Mesopotamia invented the wheel about 4000 B.C., but its close neighbour Egypt made no use of wheels for another 2,000 years. The Classic Period Maya, a literate civilization rivalling classical Europe in mathematics and astronomy, made so little use of metals that they were technically in the Stone Age. By contrast, sub-Saharan Africa mastered ironworking by 500 B.C. (as early as China did), yet never developed a full-blown civilization. The Incas of Peru, where metalworking had begun about 1500 B.C., created one of the world's largest and most closely administered empires, yet may have done so without writing as we know it (though evidence is growing that their quipu system was indeed a form of script). Japan made pottery long before anyone else—more than 12,000 years ago—but rice farming and full civilization did not appear there for another 10,000 years, adopted wholesale from China and Korea. The Japanese didn't begin to work bronze until 500 B.C., but became famous for steel swords by the sixteenth century. At that time they acquired European firearms, then abandoned them for 300 years.

We should therefore be wary of technological determinism, for it tends to underestimate cultural factors and reduce complex questions of human adaptation to a simplistic "We're the winners of history, so why didn't others do what we did?" We call agriculture and civilization "inventions" or "experiments" because that is how they look to hindsight. But they began accidentally, a series of seductive steps down a path leading, for most people, to lives of monotony and toil. Farming achieved quantity at the expense of quality: more food and more people, but seldom better nourishment or better lives. People gave up a broad array of wild foods for a handful of starchy roots and grasses—wheat, barley, rice, potatoes, maize. As we domesticated plants, the plants domesticated us. Without us, they die; and without them, so do we. There is no escape from agriculture except into mass starvation, and it has often led there anyway, with drought and blight. Most people, throughout most of time, have lived on the edge of hunger—and much of the world still does.

In hunter–gatherer societies (barring a few special cases) the social structure was more or less egalitarian, with only slight differences in wealth and power between greatest and least. Leadership was either diffuse, a matter of consensus, or something earned by merit and example. The successful hunter did not sit down beside his kill and stuff himself on the spot; he shared the meat and thereby gained prestige. If a leader became overbearing, or a minority disliked a majority decision, people could leave. In an uncrowded world without fixed borders or belongings, it was easy to vote with one's feet.

The early towns and villages that sprang up in a dozen farming heartlands around the world after the last ice age seem to have continued these free-and-easy ways for a while. Most of them were small peasant communities in which everyone worked at similar tasks and had a comparable standard of living. Land was either communally owned or thought of as having no owner but the gods. Farmers whose effort and skill made them wealthier had an obligation to share with the needy, to whom they were bound by kinship.

Gradually, however, differences in wealth and power became entrenched. Freedom and social opportunity declined as populations rose and boundaries hardened between groups. This pattern first appears in the Neolithic villages of the Middle East, and it has recurred all over the world. The first farmers along the Danube, for example, left only tools in their remains; later settlements are heavily fortified and strewn with weapons. Here, said the great Australian archaeologist Gordon Childe, "we almost see the state of war of all against all arising as … land became scarce." Writing those words in 1942, during Hitler's expansionist policy of *Lebensraum,* Childe did not need to underline how little the world had changed from Stone Age times to his.

Patriotism may indeed be, as Dr. Johnson said, "the last refuge of a scoundrel," but it's also the tyrant's first resort. People afraid of outsiders are easily manipulated. The warrior caste, supposedly society's protectors, often become protection racketeers. In times of war or crisis, power is easily stolen from the many by the few on a promise of security. The more elusive or imaginary the foe, the better for manufacturing consent. The Inquisition did a roaring trade against the Devil. And the twentieth century's struggle between capitalism and communism had all the hallmarks of the old religious wars. Was defending either system *really* worth the risk of blowing up the world?

Now we are losing hard-won freedoms on the pretext of a worldwide "war on terror," as if terrorism were something new. (Those who think it is should read *The Secret Agent,* a novel in which anarchist suicide bombers prowl London wearing explosives; it was written by Joseph Conrad a hundred years ago.) The Muslim fanatic is proving a worthy replacement for the heretic, the anarchist, and especially the Red Menace so helpful to military budgets throughout the Cold War.

Fool's Paradise

The earliest [civilization] of all was Sumer, in what is now southern Iraq. The Sumerians, whose own ethnic and linguistic stock is unclear, set a pattern that Semitic cultures and others in the Old World would follow. They came to exemplify both the best and worst of the civilized life, and they told us about themselves in cuneiform script on clay tablets, one of the most enduring mediums for the human voice, a writing like the tracks of trained birds. They set down the oldest written stories in the world, a body of texts known as *The Epic of Gilgamesh*, compiled in "strong-walled Uruk, the city of great streets" around the time that Stonehenge and the first Egyptian pyramids were being built. Legends we know from the Hebrew Bible—the Garden of Eden, the Flood—appear in Gilgamesh in earlier forms, along with other tales deemed too racy, perhaps, for inclusion in the Pentateuch. One of these, the story of the wild-man Enkidu, who is seduced into the city by "a harlot, a child of pleasure," recalls our transition from the hunting to the urban life:

> And now the wild creatures had all fled away; Enkidu was grown weak, for wisdom was in him, and the thoughts of a man were in his heart. So he returned and sat down at the woman's feet, and listened intently to what she said. "You are wise, Enkidu, and now you have become like a god. Why do you want to run wild with the beasts in the hills? Come with me. I will take you to strong-walled Uruk, to the blessed temple of Ishtar and of Anu, of love and heaven: there lives [King] Gilgamesh, who is very strong [and who] lords it over men."

In the last chapter, we left the Middle East soon after farming began in the lands often called the Fertile Crescent. Throughout human time this has been the crossroads of Africa, Europe, and Asia. Back in the Old Stone Age, Neanderthals and Cro-Magnons had contested this turf for 50,000 years—moving north and south with fluctuations in the climate, living at different times in the same rockshelters, possibly evicting one another. I suspect that if we could tune into the Middle Eastern news

at almost any period in prehistory, we would find the place seething with creativity and strife, as it has since history began.

But it's a mistake to assume that the Fertile Crescent, for all its natural endowments, its plants and animals suitable for domestication, developed quickly or easily. Even after several thousand years of farming and herding, the biggest Middle Eastern settlements—Jericho (near the Dead Sea) and Çatal Hüyük (in Anatolia)—were still tiny, covering only ten acres and thirty acres, respectively.

Insofar as the Garden of Eden had a physical geography, this was it. The serpent, however, was not the only enemy. Fortifications at Jericho and elsewhere speak of competition for land and a heavier human presence than the sites alone attest. Nor was the farming life easier or healthier than the hunting life had been: people were smaller in build and worked longer hours than non-farmers. Average life expectancy, deduced from burials at Çatal Hüyük, was twenty-nine years for women and thirty-four for men. By 6000 B.C., there is evidence of widespread deforestation and erosion. Cavalier firesetting and overgrazing by goats may have been chief culprits, but lime-burning for plaster and whitewash also destroyed the woodland, until it became the thorny scrub and semi-desert seen there today. By 5500 B.C., many of the early Neolithic sites were abandoned. As on Easter Island, people had befouled their nest, or rather had stripped it bare. But unlike the Easter Islanders, these people had room to flee and start again.

Self-driven from Eden (God's flaming sword being perhaps a glint of the fires they had set in the hills), they found a second paradise lower down on the great floodplain of the Tigris and Euphrates, the land called Mesopotamia, or Iraq. The look of this place is fresh in our minds from modern wars: treeless plains and dying oases, salt pans, dust storms, oil slicks, and burnt-out tanks. Here and there, crumbling in the ruthless sun and wind, are great mounds of mud brick—ruins of ancient cities whose names still echo in the cellars of our culture—Babylon, Uruk, and Ur of the Chaldees, where Abraham was born.

Back in the fifth and fourth millennia B.C., southern Iraq had been a marshy delta of channels teeming with fish, reeds taller than a house, and sandbars rich in date palms. Wild boar and waterfowl lived in the canebrakes. The alluvial earth, if tilled, could yield a hundredfold on every seed, for this was new land, laid down at the head of the Persian Gulf. "New" in a manner of speaking: the people who settled here had in effect followed their old fields, which had been washed from the worn hills by the great rivers flowing, as the Bible says, out of Eden.

God had spread a second chance before the children of Adam and Eve, but in this recycled Eden, unlike the first, they would eat only by sweat and toil. "The exploitation of this natural paradise," wrote Gordon Childe in his classic work, *The Most Ancient East,* "required intensive labour and the organized co-operation of large bodies of men. Arable land had literally to be created … by a 'separation' of land from water; the swamps must be drained; the floods controlled; the life-giving waters led to the rainless desert by artificial canals." It seems that in this case at least, the hierarchies of civilization grew with the demands of water control.

The scattered mud villages grew into towns. And by 3000 B.C., these towns had become small cities, rebuilt again and again on their own debris until they rose above the plain in earthen mounds known as tells. Throughout most of its thousand-year run, Sumerian civilization was dominated by a dozen such cities, each the heart of a small state. Only twice was a unified kingdom briefly forged: first by the Semitic invader Sargon, and later by the Third Dynasty of Ur. It is thought that four-fifths of the Sumerian population lived in urban centres, and

that the entire population was only half a million. (Contemporary Egypt's population was more rural and about three times this size.)

In the early days, Sumerian land was owned communally, and people brought their crops, or at least their surplus, to the city shrine, where a priesthood looked after human and divine affairs—watching the stars, directing irrigation works, improving the crops, brewing and winemaking, and building ever-grander temples. As time went by, the cities grew layer by layer into man-made hills crowned with the typical Mesopotamian step-pyramid, or ziggurat, a sacred mountain commanding the human realm. Such were the buildings the Israelites later lampooned as the Tower of Babel. The priesthoods, which had started as village co-operatives, also grew vertically to become the first corporations, complete with officials and employees, undertaking "the not unprofitable task of administering the gods' estates."

The plains of southern Iraq were rich farmland but lacked most other things town life required. Timber, flint, obsidian, metals, and every block of stone for building, carving, and food-grinding had to be imported, in return for grain and cloth. So wheeled carts, yoked oxen, and use of copper and bronze developed early. Trade and property became highly important, and have been close to the heart of Western culture ever since. Middle Easterners took a mercenary view of their gods as big landowners and themselves as serfs, "toiling in the Lord's vineyard." Unlike the writing of Egypt, China, or Mesoamerica, Sumer's writing was invented not for sacred texts, divination, literature, or even kingly propaganda, but for accounting.

Over time, the priestly corporations grew bloated and exploitive, concerned more with their own good than that of their lowlier members. Though they developed elements of capitalism, such as private ownership, there was no free competition of the kind Adam Smith recommended. The Sumerian corporations were monopolies legitimized by heaven, somewhat like mediaeval monasteries or the fiefdoms of televangelists. Their way of life, however, was far from monastic, as the temple harlotry in *Gilgamesh* implies. The Sumerian priests may have been sincere believers in their gods, though ancient people were not exempt from manipulations of credulity; at their worst, they were the world's first racketeers, running the eternal money-spinners—protection, booze, and girls.

The protection initially offered by the priesthood was from the forces of nature and the wrath of the gods. But as the Sumerian city-states grew, they began to make war among themselves. Their wealth also drew raids from mountain and desert folk, who, though less civilized, were often better armed. So it was that Uruk—at 1,100 acres and 50,000 people by far the biggest Sumerian city—became "strong-walled," the wonder of its world.

"Climb upon the wall of Uruk," invites *Gilgamesh*; "walk along it, I say; regard the foundation terrace and examine the masonry: is it not burnt brick and good?"

Having invented irrigation, the city, the corporation, and writing, Sumer added professional soldiers and hereditary kings. The kings moved out of the temples and into palaces of their own, where they forged personal links with divinity, claiming godly status by virtue of descent from heaven, a notion that would appear in many cultures and endure into modern times as divine right. With kingship came new uses for writing: dynastic history and propaganda, the exaltation of a single individual. As Bertolt Brecht dryly reflected in his poem about a worker looking at the Pyramids:

> The books are filled with names of kings.
> Was it kings who hauled the craggy blocks
> of stone? …
> Young Alexander conquered India.
> He alone?

By 2500 B.C., the days of collective landholding by city and corporation were gone; the fields now belonged to lords and great families. The Sumerian populace became serfs and sharecroppers, and beneath them was a permanent underclass of slaves—a feature of Western civilization that would last until the nineteenth century after Christ.

States arrogate to themselves the power of coercive violence: the right to crack the whip, execute prisoners, send young men to the battlefield. From this stems that venomous bloom which J. M. Coetzee has called, in his extraordinary novel *Waiting for the Barbarians,* "the black flower of civilization"—torture, wrongful imprisonment, violence for display—the forging of might into right.

Among the privileges of god-kings in Sumer and elsewhere were various styles of human sacrifice, including the right to take people along for company beyond the grave. The King's Tomb at Ur, known to archaeologists as the Death Pit, contains the first mass burial of royal concubines, retainers, and the workers who built it—about seventy-five men and women all told, their skeletons nested like spoons in a drawer. Around the world, from Egypt and Greece to China and Mexico, the idea that the king's life was worth so much more than other people's would take root again and again. The builders who seal the tomb are killed on the spot by guards, who are themselves killed by other guards, and so on, until the late king's executors deem his resting place sufficiently honoured and secure.

Since we tend to regard ancient North America as non-urban and libertarian, one of the most surprising instances of servant burial comes from Cahokia—a pre-Columbian city about the same size as Uruk—whose earthen pyramids still stand beside the Mississippi, near St. Louis.

Throughout the ancient world, rulers performed the ultimate political theatre: public sacrifice of captives. As a nineteenth-century Ashanti king candidly told the British: "If I were to abolish human sacrifice, I should deprive myself of one of the most effectual means of keeping the people in subjection." The British, who at that time were tying Indian mutineers across the mouths of cannon and blowing them in half, scarcely needed such advice. Each culture has its codes and sensibilities. In Mexico, the Spanish conquistadores were appalled by the ritual slaughter of prisoners, done with a blade to the heart. The Aztecs were equally horrified when they saw the Spaniards burn people alive.

Violence is as old as man, but civilizations commit it with a deliberation that lends it special horror. In the Death Pit of Ur we can foreglimpse all the mass graves to come, down through 5,000 years to Bosnia and Rwanda and full circle to the Iraq of Saddam Hussein, who, like the ancient kings of that land, had his name stamped on the bricks used to rebuild their monuments. In civilization, unlike the hunter-gatherer life, it has always mattered who you are. We have come a very long way from extended families around an Old Stone Age campfire to societies in which some people are demigods and others nothing more than flesh to be worked to death or buried in their betters' tombs.

Until mechanized farming began, food growers, whether peasants or slaves, outnumbered the elite and professionals who lived off their surplus by about ten to one. The masses' reward for this was usually little more than bare survival, alleviated by the consolations of custom and belief. If they were lucky, they belonged to a state that, in enlightened self-interest, would give public assistance in times of crop failure. The ideal of the leader as provider, and the wealthy as open-handed, survived to some extent and can be traced in many languages. Our word "lord" comes from the Old English *hlaford,* or "loaf-ward," he who guarded the bread supply—and was expected to share it. The Inca title *qhapaq* meant "munificent," someone who gathers wealth in but also redistributes it. Another title of the Inca emperor was *wakchakuyaq,* "he who cares for the

bereft." The chiefs of Hawaii were warned by their elders against hoarding food or goods: "The hands of the Arii must always be open; on [this] rests your prestige." And it was said of the Chinese emperors that their first duty was to feed their people. The truth is that China, like most agrarian societies, lurched from famine to famine well into modern times. Effective food security was as rare in the past as it is today in the Third World. Most ancient states did not have the storage capacity or transport to deal with anything worse than a minor crisis. The Incas and Romans were probably the best at famine relief, and it's no coincidence that both were very large empires spread over several climatic zones, with good warehousing, roads, and sea lanes.

A small civilization such as Sumer, dependent on a single ecosystem and without high ground, was especially vulnerable to flood and drought. Such disasters were viewed, then as now, as "acts of God" (or gods). Like us, the Sumerians were only dimly aware that human activity was also to blame. Floodplains will always flood, sooner or later, but deforestation of the great watersheds upstream made inundations much fiercer and more deadly than they would otherwise have been. Woodlands, with their carpet of undergrowth, mosses, and loam, work like great sponges, soaking up rainfall and allowing it to filter slowly into the earth below; trees drink up water and breathe it into the air. But wherever primaeval woods and their soils have been destroyed by cutting, burning, overgrazing, or ploughing, the bare subsoil bakes hard in dry weather and acts like a roof in wet. The result is flash floods, sometimes carrying such heavy loads of silt and gravel that they rush from steep ravines like liquid concrete. Once the waters reach a floodplain, they slow down, dump their gravel, and spread out in a brown tide that oozes its way to the sea.

Staggering alluvial forces are at work in Mesopotamia. In the 5,000 years since Sumerian records began, the twin rivers have filled in eighty miles of the Persian Gulf. Iraq's second city of Basra was open sea in ancient times. The plains of Sumer are more than two hundred miles wide. In times of an unusually great flood—the kind that might happen once a century or so—a king standing in the rain on a temple softening under his feet would see nothing but water between himself and the rim of the sky.

Not only did Adam and Eve drive themselves from Eden, but the eroded landscape they left behind set the stage for Noah's Flood. In the early days, when the city-mounds were low and easily swamped, the only refuge would have been a boat. The Sumerian version of the legend, told in the first person by a man named Utnapishtim, has the ring of real events, with vivid detail on freak weather and broken dams. In it we may see not only the forerunner of the biblical story but the first eyewitness account of a man-made environmental catastrophe:

> In those days the world teemed, the people multiplied. … Enlil heard the clamour and he said to the gods in council,
>
> "The uproar of mankind is intolerable and sleep is no longer possible. …" So the gods agreed to exterminate mankind.

Enlil, the storm god, is the instigator; others, including Ishtar, goddess of love and queen of heaven (a less virginal forerunner of Mary), go along. But Ea, the god of wisdom, warns Utnapishtim in a dream: "Tear down your house, I say, and build a boat, abandon possessions and look for life. … Take up into the boat the seed of all living creatures."

> The time was fulfilled, the evening came, the rider of the storm sent down the rain. I looked out at the weather and it was terrible, so I too boarded the boat and battened her down. … With the first light of dawn a black cloud came from the horizon; it thundered within where Adad, lord of the storm, was riding. … Then the gods of the abyss rose up; Nergal pulled out the dams of the nether waters, Ninurta the warlord threw down the dykes, and … the god of the storm turned daylight to darkness, when he smashed the land like a cup. …
>
> For six days and nights the winds blew, torrent and tempest and flood overwhelmed the world. … When the seventh day dawned … I looked at the face of the world and there was silence, all mankind was turned to clay. The surface of the sea stretched as flat as a rooftop; I opened a hatch and the light fell on my face. Then I bowed down low, I sat down and I wept … for on every side was the waste of water.

Utnapishtim sends out birds to find land. When the waters start to go down, he burns incense to draw down the gods, but his wording hints that the real attraction is the stench of corpses in the mud: the gods, he says, "gathered like flies over the sacrifice." Unlike Jehovah with his rainbow, the Sumerian deities make no promises. Ishtar fingers her necklace and says only that she will remember. Enlil sees the ark and gets angry: "Has any of these mortals escaped? Not one was to have survived." Then Ea, who had given the warning and saved the animals, upbraids Enlil for what he has done and begins a doleful chant:

> Would that a lion had ravaged mankind
> Rather than the flood. …
> Would that famine had wasted the world
> Rather than the flood.

Ea should have been more careful what he wished for. When Sir Leonard Woolley excavated in Sumer between the world wars, he wrote: "To those who have seen the Mesopotamian desert ... the ancient world seem[s] well-nigh incredible, so complete is the contrast between past and present. ... Why, if Ur was an empire's capital, if Sumer was once a vast granary, has the population dwindled to nothing, the very soil lost its virtue?"

His question had a one-word answer: salt. Rivers rinse salt from rocks and earth and carry it to the sea. But when people divert water onto arid land, much of it evaporates and the salt stays behind. Irrigation also causes waterlogging, allowing brackish groundwater to seep upward. Unless there is good drainage, long fallowing, and enough rainfall to flush the land, irrigation schemes are future salt pans.

Southern Iraq was one of the most inviting areas to begin irrigation, and one of the hardest in which to sustain it: one of the most seductive traps ever laid by progress. After a few centuries of bumper yields, the land began to turn against its tillers. The first sign of trouble was a decline in wheat, a crop that behaves like the coalminer's canary. As time went by, the Sumerians had to replace wheat with barley, which has a higher tolerance for salt. By 2500 B.C. wheat was only 15 per cent of the crop, and by 2100 B.C. Ur had given up wheat altogether.

As builders of the world's first great watering schemes, the Sumerians can hardly be blamed for failing to foresee their new technology's consequences. But political and cultural pressures certainly made matters worse. When populations were smaller, the cities had been able to sidestep the problem by lengthening fallow periods, abandoning ruined fields, and bringing new land under production, albeit with rising effort and cost. After the mid-third millennium, there was no new land to be had. Population was then at a peak, the ruling class top-heavy, and chronic warfare required the support of standing armies—nearly always a sign, and a cause, of trouble. Like the Easter Islanders, the Sumerians failed to reform their society to reduce its environmental impact. On the contrary, they tried to intensify production, especially during the Akkadian empire (c. 2350–2150 B.C.) and their swan song under the Third Dynasty of Ur, which fell in 2000 B.C.

The short-lived Empire of Ur exhibits the same behaviour as we saw on Easter Island: sticking to entrenched beliefs and practices, robbing the future to pay the present, spending the last reserves of natural capital on a reckless binge of excessive wealth and glory. Canals were lengthened, fallow periods reduced, population increased, and the economic surplus concentrated on Ur itself to support grandiose building projects. The result was a few generations of prosperity (for the rulers), followed by a collapse from which southern Mesopotamia has never recovered.

By 2000 B.C., scribes were reporting that the earth had "turned white." All crops, including barley, were failing. Yields fell to a third of their original levels. The Sumerians' thousand years in the sun of history came to an end. Political power shifted north to Babylon and Assyria, and much later, under Islam, to Baghdad. Northern Mesopotamia is better drained than the south, but even there the same cycle of degradation would be repeated by empire after empire, down to modern times. No one, it seems, was willing to learn from the past. Today, fully half of Iraq's irrigated land is saline—the highest proportion in the world, followed by the other two centres of floodplain civilization, Egypt and Pakistan.

As for the ancient cities of Sumer, a few struggled on as villages, but most were utterly abandoned. Even after 4,000 years, the land around them remains sour and barren, still white with the dust of progress. The desert in which Ur and Uruk stand is a desert of their making.

Collapse

How Societies Choose to Fail or Succeed

By Jared Diamond

Introduction

The title of Jared Diamond's *Collapse: How Societies Choose to Fail or Succeed* is quite suggestive. Much of the book is a catalog of various societies that "chose to fail." There are many such stories, beginning, as we have seen in *A Brief History of Progress*, with the ancient Sumerians. Diamond covers Easter Island, the Mayans, the Greenland Norse, and other iconic collapse cases. Diamond does not believe that human-caused environmental degradation is the sole cause of collapse; rather, he conclusively demonstrates that it is a cause among others—and the only cause within our control.

To augment this compilation's theme that we actually HAVE a global future, I have chosen to include a chapter from *Collapse* about some notable successes. From the tiny Polynesian island of Tikopia to the New Guinea highlands (examples of what Diamond calls the "bottom-up" approach) to Japan under the Tokugawas ("top-down"), societies faced with collapse have chosen to steward their resources more wisely, and have thus achieved sustainability. They did not just wake up one day and say "Let's be sustainable!" Diamond demonstrates that, in each case, only extreme difficulty—perhaps even imminent collapse—was sufficient motivation for societal change. Note too that democratic values—present in New Guinea, absent in Japan during the period in question—can contribute to or detract from ecological sustainability.

The final Diamond reading comprises two portions of the final chapter of *Collapse*: An elaboration of the 12 factors that could cause modern global society to collapse, and a final section on reasons for hope that we may succeed in the most important of human ventures: that of creating a sustainable global society. The "12 factors" analysis amazes me by condensing the human impact information of an entire environmental science course into a few pages. As for the "hope" section, my take-home from Diamond is this: *Other* societies have been able to transition from crisis to sustainability: with sufficient political will, so can *we*.

[Diamond begins this chapter by delineating between "bottom-up" and "top-down" approaches to sustainability. Bottom-up approaches are typical to small societies where everyone knows everyone else's business and the effects of environmental destruction are local and plain to see. United by a common interest, such societies more easily cooperate. Diamond gives a neighborhood association as an example of a successful bottom-up cooperative approach the reader may be familiar with. In this chapter, he uses a small island (Tikopia) and a "primitive" society (highland New Guinea) as examples of such bottom-up sustainability.

The other approach is "top-down" sustainability, of which Diamond's example is Japan under the Tokugawa regime. Top-down approaches work with large centralized societies. The top leadership is capable of perceiving systemic problems across the whole society that locals may be unable to perceive or simply unable to respond to. If the leadership has a motivation for wanting the realm to be sustainable (such as ensuring succession or re-election) then top-down approaches are rewarded and may be successful. Of course many societies, such as our own federal system which shares power between local, state and federal governments, are a mixture of bottom-up and top-down approaches. Medium-sized societies may be less successful; Diamond discusses the island of Mangaia as an example. "Hostility between chiefs in neighborhood valleys prevents agreement or coordinated action, and even contributes to environmental destruction: each chief leads raids to cut down trees and wreak havoc on rivals' land. The island may be too small for a central government to have arisen, capable of controlling the whole island."]

The first example is the highlands of New Guinea, one of the world's great success stories of bottom-up management. People have been living self-sustainably in New Guinea for about 46,000 years, until recent times without economically significant inputs from societies outside the highlands, and without inputs of any sort except trade items prized just for status (such as cowry shells and bird-of-paradise plumes). New Guinea is the large island just north of Australia (map, p. 84), lying almost on the equator and hence with hot tropical rainforest in the lowlands, but whose rugged interior consists of alternating ridges and valleys culminating in glacier-covered mountains up to 16,500 feet high. The terrain ruggedness confined European explorers to the coast and lowland rivers for 400 years, during which it became assumed that the interior was forest-covered and uninhabited.

It was therefore a shock, when airplanes chartered by biologists and miners first flew over the interior in the 1930s, for the pilots to see below them a landscape transformed by millions of people previously unknown to the outside world. The scene looked like the most densely populated areas of Holland (Plate 19): broad open valleys with few clumps of trees, divided as far as the eye could see into neatly laid-out gardens separated by ditches for irrigation and drainage, terraced steep hillsides reminiscent of Java or Japan, and villages surrounded by defensive stockades. When more Europeans followed up the pilots' discoveries overland, they found that the inhabitants were farmers who grew taro, bananas, yams, sugarcane, sweet potatoes, pigs, and chickens. We now know that the first four of those major crops (plus other minor ones) were domesticated in New Guinea itself, that the New Guinea highlands were one of only nine independent centers of plant domestication in the world, and that agriculture has been going on there for about 7,000 years—one of the world's longest-running experiments in sustainable food production.

To European explorers and colonizers, New Guinea highlanders seemed "primitive." They lived in thatched huts, were chronically at war with each other, had no kings or even chiefs, lacked writing, and wore little or no clothing even under cold conditions with heavy rain. They lacked metal and made their tools instead of stone, wood, and bone. For instance, they felled trees with stone axes, dug gardens and ditches with wooden sticks, and fought each other with wooden spears and arrows and bamboo knives.

That "primitive" appearance proved deceptive, because their farming methods are sophisticated, so much so that European agronomists still don't understand today in some cases the reasons why New Guineans' methods work and why well-intentioned European farming innovation failed there. For instance, one European agricultural advisor was horrified to notice that a New Guinean sweet potato garden on a steep slope in a wet area had vertical drainage ditches running straight down the slope. He convinced the villagers to correct their awful mistake, and instead to put in drains running horizontally along contours, according to good European practices. Awed by him, the villagers reoriented their drains, with the result that water built up behind the drains, and in the next heavy rains a landslide carried the entire garden down the slope into the river below. To avoid exactly that outcome, New Guinea farmers long before the arrival of Europeans learned the virtues of vertical drains under highland rain and soil conditions.

That's only one of the techniques that New Guineans worked out by trial and error, over the course of thousands of years, for growing crops in areas receiving up to 400 inches of rain per year, with frequent earthquakes, landslides, and (at higher elevations) frost. To maintain soil fertility, especially in areas of high population density where short fallow periods or even continuous growing of crops were essential to produce enough food, they resorted to a whole suite of techniques besides the silviculture that I'll explain in a moment. They added weeds, grass, old vines, and other organic matter to the soil as compost at up to 16 tons per acre. They applied garbage, ash from fires, vegetation cut from fields resting in fallow, rotten logs, and chicken manure as mulches and fertilizers to the soil surface. They dug ditches around fields to lower the watertable and prevent waterlogging, and transferred the organic muck dug out of those ditches onto the soil surface. Legume food crops that fix atmospheric nitrogen, such as beans, were rotated with other crops-in effect, an independent New Guinean invention of a crop rotation principle now widespread in First World agriculture for maintaining soil nitrogen levels. On steep slopes New Guineans constructed terraces, erected soil retention barriers, and of course removed excess water by the vertical drains that aroused the agronomist's ire. A consequence of their relying on all these specialized methods is that it takes years of growing up in a village to learn how to farm successfully in the New Guinea highlands. My highland friends who spent their childhood years away from their village to pursue an education found, on returning to the village, that they were incompetent at farming their family gardens because they had missed out on mastering a large body of complex knowledge.

Sustainable agriculture in the New Guinea highlands poses difficult problems not only of soil fertility but also of wood supplies, as a result of forests having to be cleared for gardens and villages. The traditional highland lifestyle relied on trees for many purposes, such as for timber to build houses and fences, wood for making tools and utensils and weapons, and fuel for cooking and for heating the hut during the cold nights. Originally, the highlands were covered with oak and beech forests, but thousands of years of gardening have left the most densely populated areas (especially the Wahgi Valley of Papua New Guinea and the Baliem Valley of Indonesian New Guinea) completely deforested up to an elevation of 8,000 feet. Where do highlanders obtain all the wood that they need?

Already on the first day of my visit to the highlands in 1964, I saw groves of a species of casuarina tree in villages and gardens. Also known as she oaks or ironwood, casuarinas are a group of several dozen tree species with leaves resembling pine needles, native to Pacific islands, Australia, Southeast Asia, and tropical East Africa, but now widely introduced elsewhere because of their easily split but very hard wood (hence that name "ironwood"). A species native to the New Guinea highlands, *Casuarina oligodon,* is the one that several million highlanders grow on a massive scale by transplanting seedlings that have sprouted naturally along stream banks. Highlanders similarly plant several other tree species, but casuarina is the most prevalent. So extensive is the scale of transplanting casuarinas in the highlands that the practice is now referred to as "silviculture," the growing of trees instead of field crops as in conventional agriculture (*silva, ager,* and *cultura* are the Latin words for woodland, field, and cultivation, respectively).

Only gradually have European foresters come to appreciate the particular advantages of *Casuarina oligodon,* and the benefits that highlanders obtain from its groves. The species is fast-growing. Its wood is excellent for timber and fuel. Its root nodules that fix nitrogen, and its copious leaf-fall, add both nitrogen and carbon to the soil. Hence casuarinas grown interspersed in active gardens increase the soil's fertility, while casuarinas grown in abandoned gardens shorten the length of time that the site must be left fallow to recover its fertility before a new crop can be planted. The roots hold soil on steep slopes and thereby reduce erosion, New Guinea

farmers claim that the trees somehow reduce garden infestation with a taro beetle, and experience suggests that they are right about that claim as they are about many others, though agronomists still haven't figured out the basis of the tree's claimed anti-beetle potency. Highlanders also say that they appreciate their casuarina groves for esthetic reasons, because they like the sound of the wind blowing through the branches, and because the trees provide shade to the village. Thus, even in broad valleys from which the original forest has been completely cleared, casuarina silviculture permits a wood-dependent society to continue to thrive.

How long have New Guinea highlanders been practicing silviculture? The clues used by paleobotanists to reconstruct the vegetational history of the highlands have been basically similar to those I already discussed for Easter Island, the Maya area, Iceland, and Greenland in Chapters 2–8: analysis of swamp and lake cores for pollen identified down to the level of the plant species producing the pollen; presence of charcoal or carbonized particles resulting from fires (either natural or else lit by humans to clear forests); sediment accumulation suggesting erosion following forest clearance; and radiocarbon dating.

It turns out that New Guinea and Australia were first settled around 46,000 years ago by humans moving eastwards from Asia through Indonesia's islands on rafts or canoes. At that time, New Guinea was still joined in a single landmass to Australia, where early human arrival is well attested at numerous sites. By 32,000 years ago, the appearance of charcoal from frequent fires and an increase in pollen of non-forest tree species compared to forest tree species at New Guinea highland sites hint that people were already visiting the sites, presumably to hunt and to gather forest pandanus nuts as they still do today. Signs of sustained forest clearance and the appearance of artificial drains within valley swamps by around 7,000 years ago suggest the origins of highland agriculture then. Forest pollen continues to decrease at the expense of non-forest pollen until around 1,200 years ago, when the first big surge in quantities of casuarina pollen appears almost simultaneously in two valleys 500 miles apart, the Baliem Valley in the west and the Wahgi Valley in the east. Today those are the broadest, most extensively deforested highland valleys, supporting the largest and densest human populations, and those same features were probably true of those two valleys 1,200 years ago.

If we take that casuarina pollen surge as a sign of the beginning of casuarina silviculture, why should it have arisen then, apparently independently in two separate areas of the highlands? Two or three factors were working together at that time to produce a wood crisis. One was the advance of deforestation, as the highland's farming population increased from 7,000 years ago onwards. A second factor is associated with a thick layer of volcanic ashfall, termed the Ogowila tephra, which at just that time blanketed eastern New Guinea (including the Wahgi Valley) but wasn't blown as far west as the Baliem Valley. That Ogowila tephra originated from an enormous eruption on Long Island off the coast of eastern New Guinea. When I visited Long Island in 1972, the island consisted of a ring of mountains 16 miles in diameter surrounding a huge hole filled by a crater lake, one of the largest lakes on any Pacific island. As discussed in Chapter 2, the nutrients carried in such an ashfall would have stimulated crop growth and thereby stimulated human population growth, in turn creating increased need for wood for timber and fuel, and increased rewards for discovering the virtues of casuarina silviculture. Finally, if one can extrapolate to New Guinea from the time record of El Niño events demonstrated for Peru, droughts and frost might have stressed highland societies then as a third factor.

To judge by an even bigger surge in casuarina pollen between 300 and 600 years ago, highlanders may then have expanded silviculture further under the stimulus of two other events: the Tibito tephra, an even bigger volcanic ashfall and boost to soil fertility and human population than the Ogowila tephra, also originating from Long Island and directly responsible for the hole filled by the modern lake that I saw; and possibly the arrival then of the Andean sweet potato in the New Guinea highlands, permitting crop yields several times those previously available with just New Guinean crops. After its initial appearance in the Wahgi and Baliem Valleys, casuarina silviculture (as attested by pollen cores) reached other highland areas at various later times, and was adopted in some outlying areas only within the 20th century. That spread of silviculture probably involved diffusion of knowledge of the technique from its first two sites of invention, plus perhaps some later independent inventions in other areas.

I have presented New Guinea highland casuarina silviculture as an example of bottom-up problem-solving, even though there are no written records from the highlands to tell us exactly how the technique was adopted. But it could hardly have been by any other type of problem-solving, because New Guinea highland societies represent an ultra-democratic extreme of bottom-up decision-making. colonial government in the 1930s, there had not been even any beginnings of political unification in any part of the highlands: merely individual villages alternating between fighting each other and joining in temporary alliances with each other against other nearby villages. Within each village, instead of hereditary leaders or chiefs, there were just individuals, called "big-men," who by force of personality were more influential than other individuals but still lived in a hut like everybody else's and tilled a garden like anybody else's. Decisions were (and often still are today) reached by means of everybody in the village sitting down together and talking, and talking, and talking. The big-men couldn't give orders, and they might or might not succeed in persuading others to adopt their proposals. To outsiders today (including not just me but often New Guinea government officials themselves), that bottom-up approach to decision-making can be frustrating, because you can't go to some designated village leader and get a quick answer to your request; you have to have the patience to endure talk-talk-talk for hours or days with every villager who has some opinion to offer.

That must have been the context in which casuarina silviculture and all those other useful agricultural practices were adopted in the New Guinea highlands. People in any village could see the deforestation going on around them, could recognize the lower growth rates of their crops as gardens lost fertility after being initially cleared, and experienced the consequences of timber and fuel scarcity. New Guineans are more curious and experimental than any other people that I have encountered. When in my early years in New Guinea I saw someone who had acquired a pencil, which was still an unfamiliar object then, the pencil would be tried out for myriad purposes other than writing: a hair decoration? a stabbing tool? something to chew on? a long earring? a plug through the pierced nasal septum? Whenever I take New Guineans to work with me in areas away from their own village, they are constantly picking up local plants, asking local people about the plants' uses, and selecting some of the plants to bring back with them and try growing at home. In that way, someone 1,200 years ago would have noticed the casuarina seedlings growing beside a stream, brought them home as yet another plant to try out, noticed the beneficial effects in a garden—and then some other people would have observed those garden casuarinas and tried the seedlings for themselves.

Besides thereby solving their problems of wood supply and soil fertility, New Guinea highlanders also faced a population problem as their numbers increased. That population increase became checked by practices that continued into the childhoods of many of my New Guinea friends—especially by war, infanticide, use of forest plants for contraception and abortion, and sexual abstinence and natural lactational amenorrhea for several years while a baby was being nursed. New Guinea societies thereby avoided the fates that Easter Island, Mangareva, the Maya, the Anasazi, and many other societies suffered through deforestation and population growth. Highlanders managed to operate sustainably for tens of thousands of years before the origins of agriculture, and then for another 7,000 years after the origins of agriculture, despite climate changes and human environmental impacts constantly creating altered conditions.

Today, New Guineans are facing a new population explosion because of the success of public health measures, introduction of new crops, and the end or decrease of intertribal warfare. Population control by infanticide is no longer socially acceptable as a solution. But New Guineans already adapted in the past to such big changes as the extinction of the Pleistocene megafauna, glacial melting and warming temperatures at the end of the Ice Ages, the development of agriculture, massive deforestation, volcanic tephra fallouts, El Niño events, the arrival of the sweet potato, and the arrival of Europeans. Will they now also be able to adapt to the changed conditions producing their current population explosion?

Tikopia, a tiny, isolated, tropical island in the Southwest Pacific Ocean, is another success story of bottom-up management (map, p. 84). With a total area of just 1.8 square miles, it supports 1,200 people, which works out to a population density of 800 people per square mile of farmable land. That's a dense population for a traditional society without modern agricultural techniques. Nevertheless, the island has been occupied continuously for almost 3,000 years.

The nearest land of any sort to Tikopia is the even-tinier (one-seventh of a square mile) island of Anuta 85 miles distant, inhabited by only 170 people. The nearest larger islands, Vanua Lava and Vanikoro in the Vanuatu and Solomon Archipelagoes respectively, are 140 miles distant and still only 100 square miles each in area. In the words of the anthropologist Raymond Firth, who lived on Tikopia for a year in 1928–29 and returned for subsequent visits, "It's hard for anyone who has not actually lived on the island to realize its isolation from the rest of the world. It is so small that one is rarely out of sight or sound of the sea. [The maximum distance from the center of the island to the coast is three-quarters of a mile.] The native concept of space bears a distinct relation to this. They find it almost impossible to conceive of any really large land mass. ... I was once asked seriously by a group of them, 'Friend, is there any land where the sound of the sea is not heard?' Their confinement has another less obvious result. For all kinds of spatial reference they use the expressions *inland* and *to seawards*. Thus an axe lying on the floor of a house is localized in this way, and I have even heard a man direct the attention of another in saying: 'There is a spot of mud on your seaward cheek.' Day by day, month after month, nothing breaks the level line of a clear horizon, and there is no faint haze to tell of the existence of any other land."

In Tikopia's traditional small canoes, the open-ocean voyage over the cyclone-prone Southwest Pacific to any of those nearest-neighbor islands was dangerous, although Tikopians considered it a great adventure. The canoes' small sizes and the infrequency of the voyages severely limited the quantity of goods that could be imported, so that in practice the only economically significant imports were stone for making tools, and unmarried young people from Anuta as marriage partners. Because Tikopia rock is

of poor quality for making tools (just as we saw for Mangareva and Henderson Islands in Chapter 3), obsidian, volcanic glass, basalt, and chert were imported from Vanua Lava and Vanikoro, with some of that imported stone in turn originating from much more distant islands in the Bismarck, Solomon, and Samoan Archipelagoes. Other imports consisted of luxury goods: shells for ornaments, bows and arrows, and (formerly) pottery.

There could be no question of importing staple foods in amounts sufficient to contribute meaningfully to Tikopian subsistence. In particular, Tikopians had to produce and store enough surplus food to be able to avoid starvation during the annual dry season of May and June, and after cyclones that at unpredictable intervals destroy gardens. (Tikopia lies in the Pacific's main cyclone belt, with on the average 20 cyclones per decade.) Hence surviving on Tikopia required solving two problems for 3,000 years: How could a food supply sufficient for 1,200 people be produced reliably? And how could the population be prevented from increasing to a higher level that would be impossible to sustain?

Our main source of information about the traditional Tikopian lifestyle comes from Firth's observations, one of the classic studies of anthropology. While Tikopia had been "discovered" by Europeans already in 1606, its isolation ensured that European influence remained negligible until the 1800s, the first visit by missionaries did not take place until 1857, and the first conversions of islanders to Christianity did not begin until after 1900. Hence Firth in 1928–29 had a better opportunity than subsequent visiting anthropologists to observe a culture that still contained many of its traditional elements, although already then in the process of change.

Sustainability of food production on Tikopia is promoted by some of the environmental factors discussed in Chapter 2 as tending to make societies on some Pacific islands more sustainable, and less susceptible to environmental degradation, than societies on other islands. Working in favor of sustainability on Tikopia are its high rainfall, moderate latitude and location in the zone of high volcanic ash fallout (from volcanoes on other islands) and high fallout of Asian dust. Those factors constitute a geographical stroke of good luck for the Tikopians: favorable conditions for which they personally could claim no credit. The remainder of their good fortune must be credited to what they have done for themselves. Virtually the whole island is micromanaged for continuous and sustainable food production, instead of the slash-and-burn agriculture prevalent on many other Pacific islands. Almost every plant species on Tikopia is used by people in one way or another: even grass is used as a mulch in gardens, and wild trees are used as food sources in times of famine.

As you approach Tikopia from the sea, the island appears to be covered with tall, multi-storied, original rainforest, like that mantling uninhabited Pacific islands. Only when you land and go among the trees do you realize that true rainforest is confined to a few patches on the steepest cliffs, and that the rest of the island is devoted to food production. Most of the island's area is covered with an orchard whose tallest trees are native or introduced tree species producing edible nuts or fruit or other useful products, of which the most important are coconuts, breadfruit, and sago palms yielding a starchy pith. Less numerous but still valuable canopy trees are the native almond (*Canarium harveyi*), the nut-bearing *Burckella ovovata*, the Tahi-tian chestnut *Inocarpus fagiferus*, the cut-nut *Barringtonia procera*, and the tropical almond *Terminalia catappa*. Smaller useful trees in the middle story include the betelnut palm with narcotic-containing nuts, the vi-apple *Spondias dulcis*, and the medium-sized mami tree *Antiaris toxicara*, which fits well into this orchard and whose bark was used for cloth, instead of the paper mulberry used on other Polynesian islands. The understory below these tree layers i s i n effect a garden for

growing yams, bananas, and the giant swamp taro *Cyrtosperma chamissonis*, most of whose varieties require swampy conditions but of which Tikopians grow a genetic clone specifically adapted to dry conditions in their well-drained hillside orchards. This whole multi-story orchard is unique in the Pacific in its structural mimicry of a rainforest, except that its plants are all edible whereas most rainforest trees are inedible.

In addition to these extensive orchards, there are two other types of small areas that are open and treeless but also used for food production. One is a small freshwater swamp, devoted to growing the usual moisture-adapted form of giant swamp taro instead of the distinctive dry-adapted clone grown on hillsides. The other consists of fields devoted to short-fallow, labor-intensive, nearly continuous production of three root crops: taro, yams, and now the South American–introduced crop manioc, which has largely replaced native yams. These fields require almost constant labor input for weeding, plus mulching with grass and brushwood to prevent crop plants from drying out.

The main food products of these orchards, swamps, and fields are starchy plant foods. For their protein, in the absence of domestic animals larger than chickens and dogs, traditional Tikopians relied to a minor extent on ducks and fish obtained from the island's one brackish lake, and to a major extent on fish and shellfish from the sea. Sustainable exploitation of seafood resulted from taboos administered by chiefs, whose permission was required to catch or eat fish; the taboos therefore had the effect of preventing overfishing.

Tikopians still had to fall back on two types of emergency food supply to get them over the annual dry season when crop production was low, and the occasional cyclone that could destroy gardens and orchard crops. One type consisted of fermenting surplus breadfruit in pits to produce a starchy paste that can be stored for two or three years. The other type consisted of exploiting the small remaining stands of original rainforest to harvest fruits, nuts, and other edible plant parts that were not preferred foods but could save people from otherwise starving. In 1976, while I was visiting another Polynesian island called Rennell, I asked Rennell Islanders about the edibility of fruit from each of the dozens of Rennell species of forest trees. There proved to be three answers: some trees were said to have "edible" fruit; some trees were said to have "inedible" fruit; and other trees had fruit "eaten only at the time of the *hungi kenge.*" Never having heard of a *hungi kenge,* I inquired about it. I was told that it was the biggest cyclone in living memory, which had destroyed Rennell's gardens around 1910 and reduced people to the point of starvation, from which they saved themselves by eating forest fruits that they didn't especially like and normally wouldn't eat. On Tikopia, with its two cyclones in the average year, such fruits must be even more important than on Rennell.

———————

[Diamond goes on to note that, besides managing the environment to maximize sustainable food supply, population control was the other important part of the equation for success on Tikopia. Zero population growth is an explicit goal, and is discussed in detail and vigrously promoted by Tikopia's chiefs. Historically it was achieved by seven methods: contraception (by coitus interruptus, i.e. "pulling out"), abortion, suicide, infanticide, celibacy, reckless overseas voyages (so dangerous Diamond calls them "virtual suicide"), and murder (a rare event; there is only one instance of clan warfare in Tikopia's oral history). Of these population control methods, only contraception is widely acceptable in modern societies. The modern reader will perhaps be shocked by the latter six methods; but it is important to note that coitus interruptus often fails to prevent pregnancy. In the absence of modern contraceptive techniques,

many traditional societies have resorted to the latter six methods. We may judge them for this, but if we do so, we must face a grim question. What's better: abortion, infanticide and murder, or total societal collapse, where everyone could die?]

———————

[Tikopia's sustainable society was] not invented all at once but developed over the course of nearly 3,000 years. The island was first settled around 900 B.C. by Lapita people ancestral to the modern Polynesians, as described in Chapter 2. Those first settlers made a heavy impact on the island's environment. Remains of charcoal at archaeological sites show that they cleared forest by burning it. They feasted on breeding colonies of seabirds, land birds, and fruit bats, and on fish, shellfish, and sea turtles. Within a thousand years, the Tikopian populations of five bird species (Abbott's Booby, Audubon's Shearwater, Banded Rail, Common Megapode, and Sooty Tern) were extirpated, to be followed later by the Red-footed Booby. Also in that first millennium, archaeological middens reveal the virtual elimination of fruit bats, a three-fold decrease in fish and bird bones, a 10-fold decrease in shellfish, and a decrease in the maximum size of giant clams and turban shells (presumably because people were preferentially harvesting the largest individuals).

Around 100 B.C., the economy began to change as those initial food sources disappeared or were depleted. Over the course of the next thousand years, charcoal accumulation ceased, and remains of native almonds (*Canarium harveyi*) appeared, in archaeological sites, indicating that Tikopians were abandoning slash-and-burn agriculture in favor of maintaining orchards with nut trees. To compensate for the drastic declines in birds and seafood, people shifted to intensive husbandry of pigs, which came to account for nearly half of all protein consumed. An abrupt change in economy and artifacts around A.D. 1200 marks the arrival of Polynesians from the east, whose distinctive cultural features had been forming in the area of Fiji, Samoa, and Tonga among descendants of the Lapita migration that had initially also colonized Tikopia. It was those Polynesians who brought with them the technique of fermenting and storing breadfruit in pits.

A momentous decision taken consciously around A.D. 1600, and recorded in oral traditions but also attested archaeologically, was the killing of every pig on the island, to be replaced as protein sources by an increase in consumption of fish, shell-fish, and turtles. According to Tikopians' accounts, their ancestors had made that decision because pigs raided and rooted up gardens, competed with humans for food, were an inefficient means to feed humans (it takes about 10 pounds of vegetables edible to humans to produce just one pound of pork), and had become a luxury food for the chiefs. With that elimination of pigs, and the transformation of Tikopia's bay into a brackish lake around the same time, Tikopia's economy achieved essentially the form in which it existed when Europeans first began to take up residence in the 1800s. Thus, until colonial government and Christian mission influence became important in the 20th century, Tikopians had been virtually self-supporting on their micromanaged remote little speck of land for three millennia.

Tikopians today are divided among four clans each headed by a hereditary chief, who holds more power than does a non-hereditary big-man of the New Guinea highlands. Nevertheless, the evolution of Tikopian subsistence is better described by the bottom-up metaphor than by the top-down metaphor. One can walk all the way around the coastline of Tikopia in under half a day, so that every Tikopian is familiar with the entire island. The population is small enough that every Tikopian resident on the island can also know all other residents individually. While every piece of land has a name and is owned by some patrilineal kinship group, each house owns pieces of land in different parts of the island. If a garden is not being used at the moment, anyone can temporarily plant crops in that garden without asking the owner's permission. Anyone can fish on any reef, regardless of whether it happens to be in front of someone else's house. When a cyclone or drought arrives, it affects the entire island. Thus, despite differences among Tikopians in their clan affiliation and in how much land their kinship group owns, they all face the same problems and are at the mercy of the same dangers. Tikopia's isolation and small size have demanded collective decision-making ever since the island was settled. Anthropologist Raymond Firth entitled his first book *We, the Tikopia* because he often heard that phrase ("*Matou nga Tikopia*") from Tikopians explaining their society to him.

Tikopia's chiefs do serve as the overlords of clan lands and canoes, and they redistribute resources. By Polynesian standards, however, Tikopia is among the least stratified chiefdoms with the weakest chiefs. Chiefs and their families produce their own food and dig in their own gardens and orchards, as do commoners. In Firth's words, "Ultimately the mode of production is inherent in the social tradition, of which the chief is merely the prime agent and interpreter. He and his people share the same values: an ideology of kinship, ritual, and morality reinforced by legend and mythology. The chief is to a considerable extent a custodian of this tradition, but he is not alone in this. His elders, his fellow chiefs, the people of his clan, and even the members of his family are all imbued with the same values, and advise and criticize his actions." Thus, that role of Tikopian chiefs represents much less top-down management than does the role of the leaders of the remaining society that we shall now discuss.

———————

Our other success story resembles Tikopia in that it too involves a densely populated island society isolated from the outside world, with few economically significant imports, and with a long history of a self-sufficient and sustainable lifestyle. But the resemblance ends there, because this island has a population 100,000 times larger than Tikopia's, a powerful central government, an industrial First World economy, a highly stratified society presided over by a rich powerful elite, and a big role of top-down initiatives in solving environmental problems. Our case study is of Japan before 1868.

Japan's long history of scientific forest management is not well known to Europeans and Americans. Instead, professional foresters think of the techniques of forest management widespread today as having begun to develop in German principalities in the 1500s, and having spread from there to much of the rest of Europe in the 1700s and 1800s. As a result, Europe's total area of forest, after declining steadily ever since the origins of European agriculture 9,000 years ago, has actually been increasing since around 1800. When I first visited Germany in 1959, I was astonished to discover the extent of neatly laid-out forest plantations covering much of the country, because I had thought of Germany as industrialized, populous, and urban.

But it turns out that Japan, independently of and simultaneously with Germany, also developed top-down forest management. That too is surprising, because Japan, like Germany, is industrialized, populous, and urban. It has the highest population density of any large First World country, with nearly 1,000 people per square mile of total area, or 5,000 people per square mile of farmland. Despite that high population, almost 80% of Japan's area consists of sparsely populated forested mountains (Plate 20), while most people and agriculture are crammed into the plains that make up only one-fifth of the country. Those forests are so well protected and managed that their extent is still increasing, even though they are being utilized as valuable sources of timber. Because of that forest mantle, the Japanese often refer to their island nation as "the green archipelago." While the mantle superficially resembles a primeval forest, in fact most of Japan's accessible original forests were cut by 300 years ago and became replaced with regrowth forest and plantations as tightly micromanaged as those of Germany and Tikopia.

Japanese forest policies arose as a response to an environmental and population crisis paradoxically brought on by peace and prosperity. For almost 150 years beginning in 1467, Japan was convulsed by civil wars as the ruling coalition of powerful houses that had emerged from the earlier disintegration of the emperor's power in turn collapsed, and as control passed instead to dozens of autonomous warrior barons (called *daimyo*), who fought each other. The wars were finally ended by the military victories of a warrior named Toyotomi Hideyoshi and his successor Tokugawa Ieyasu. In 1615 Ieyasu's storming of the Toyotomi family stronghold at Osaka, and the deaths by suicide of the remaining Toyotomis, marked the wars' end.

Already in 1603, the emperor had invested Ieyasu with the hereditary title of *shogun,* the chief of the warrior estate. From then on, the shogun based at his capital city of Edo (modern Tokyo) exercised the real power, while the emperor at the old capital of Kyoto remained a figurehead. A quarter of Japan's area was directly administered by the shogun, the remaining three-quarters being administered by the 250 daimyo whom the shogun ruled with a firm hand. Military force became the shogun's monopoly. Daimyo could no longer fight each other, and they even needed the shogun's permission to marry, to modify their castles, or to pass on their property in inheritance to a son. The years from 1603 to 1867 in Japan are called the Tokugawa era, during which a series of Tokugawa shoguns kept Japan free of war and foreign influence.

Peace and prosperity allowed Japan's population and economy to explode. Within a century of the wars' end, population doubled because of a fortunate combination of factors: peaceful conditions, relative freedom from the disease epidemics afflicting Europe at the time (due to Japan's ban on foreign travel or visitors: see below), and increased agricultural productivity as the result of the arrival of two productive new crops (potatoes and sweet potatoes), marsh reclamation, improved flood control, and increased production of irrigated rice. While the population as a whole thus grew, cities grew even faster, to the point where Edo became the world's most populous city by 1720. Throughout Japan, peace and a strong centralized government brought a uniform currency and uniform system of weights and measures, the end of toll and customs barriers, road construction, and improved coastal shipping, all of which contributed to a trade boom within Japan.

But Japan's trade with the rest of the world was cut to almost nothing. Portuguese navigators bent on trade and conquest, having rounded Africa to reach India in 1498, advanced to the Moluccas in 1512, China in 1514, and Japan in 1543. Those first European visitors to Japan were just a pair of shipwrecked sailors, but they caused unsettling changes by introducing guns, and even bigger changes when they were followed by Catholic missionaries six years later. Hundreds of thousands of Japanese, including some daimyo, became converted to Christianity. Unfortunately, rival Jesuit and Franciscan missionaries began competing with each other, and stories spread that friars were trying to Christianize Japan as a prelude to a European takeover.

In 1597 Toyotomi Hideyoshi crucified Japan's first group of 26 Christian martyrs. When Christian daimyo then tried to bribe or assassinate government officials, the shogun Tokugawa Ieyasu concluded that Europeans and Christianity posed a threat to the stability of the shogunate and Japan. (In retrospect, when one considers how European military intervention followed the arrival of apparently innocent traders and missionaries in China, India, and many other countries, the threat foreseen by Ieyasu was real.) In 1614 Ieyasu prohibited Christianity and began to torture and execute missionaries and those of their converts who refused to disavow their religion. In 1635 a later shogun went even further by forbidding Japanese to travel overseas and forbidding Japanese ships to leave Japan's coastal waters. Four years later, he expelled all the remaining Portuguese from Japan.

Japan thereupon entered a period, lasting over two centuries, in which it cordoned itself off from the rest of the world, for reasons reflecting even more its agendas related to China and Korea than to Europe. The sole foreign traders admitted were a few Dutch merchants (considered less dangerous than Portuguese because they were anti-Catholic), kept isolated like dangerous germs on an island in Nagasaki harbor, and a similar Chinese enclave. The only other foreign trade permitted was with Koreans on Tsushima Island lying between Korea and Japan, with the Ryukyu Islands (including Okinawa) to the south, and with the aboriginal Ainu population on Hokkaido Island to the north (then not yet part of Japan, as it is today). Apart from those contacts, Japan did not even maintain overseas diplomatic relations, not even with China. Nor did Japan attempt foreign conquests after Hideyoshi's two unsuccessful invasions of Korea in the 1590s.

During those centuries of relative isolation. Japan was able to meet most of its needs domestically, and in particular was virtually self-sufficient in food, timber, and most metals. Imports were largely restricted to sugar and spices, ginseng and medicines and mercury, 160 tons per year of luxury woods, Chinese silk, deer skin and other hides to make leather (because Japan maintained few cattle), and lead and saltpeter to make gunpowder. Even the amounts of some of those imports decreased with time as domestic silk and sugar production rose, and as guns became restricted and then virtually abolished. This remarkable state of self-sufficiency and self-imposed isolation lasted until an American fleet under Commodore Perry arrived in 1853 to demand that Japan open its ports to supply fuel and provisions to American whaling and merchant ships. When it then became clear that the Tokugawa shogunate could no longer protect Japan from barbarians armed with guns, the shogunate collapsed in 1868, and Japan began its remarkably rapid transformation from an isolated semi-feudal society to a modern state.

Deforestation was a major factor in the environmental and population crisis brought on by the peace and prosperity of the 1600s, as Japan's timber consumption (almost entirely consisting of domestic timber) soared. Until the late 19th century, most Japanese buildings were made of wood, rather than of stone, brick, cement, mud, or tiles as in many other countries. That tradition of timber construction stemmed partly from a Japanese esthetic preference for wood, and partly from the ready availability of trees through-out Japan's early history. With the onset of peace, prosperity, and a population boom, timber use for construction took off to supply the needs of the growing rural and urban population. Beginning around 1570, Hideyoshi, his successor the shogun Ieyasu, and many of the daimyo led the way, in- dulging their egos and seeking to impress each other by constructing huge castles and temples. Just the three biggest castles built by Ieyasu required clear-cutting about 10 square miles of forests. About 200 castle towns and cities arose under Hideyoshi, Ieyasu, and the next shogun. After Ieyasu's death, urban construction outstripped elite monument construction in its demand for timber, especially because cities of thatch-roofed wooden buildings set closely together and with winter heating by fireplaces were prone to burn, so cities needed to be rebuilt repeatedly. The biggest of those urban fires was the Meireki fire that burned half of the capital at Edo and killed 100,000 people in 1657. Much of that timber was transported to cities by coastal ships, in turn built of wood and hence consuming more wood. Still more wooden ships were required to transport Hideyoshi's armies across the Korea Strait in his unsuccessful attempts to conquer Korea.

Timber for construction was not the only need driving deforestation. Wood was also the fuel used for heating houses, for cooking, and for industrial uses such as making salt, tiles, and ceramics. Wood was burned to charcoal to sustain the hotter fires required for smelting iron. Japan's expanding population needed more food, and hence more forested land cleared for agriculture. Peasants fertilized their fields with "green fertilizer" (i.e., leaves, bark, and twigs), and fed their oxen and horses with fodder (brush and grass), obtained from the forests. Each acre of cropland required 5 to 10 acres of forest to provide the necessary green fertilizer. Until the civil wars ended in 1615, the warring armies under daimyo and the shogun took fodder for their horses, and bamboo for their weapons and defensive palisades, from the forests. Daimyo in forested areas fulfilled their annual obligation to the shogun in the form of timber.

The years from about 1570 to 1650 marked the peak of the construction boom and of deforestation, which slowed down as timber became scarce. At first, wood was cut either under the direct order of the shogun or daimyo, or else by peasants themselves for their local needs, but by 1660 logging by private entrepreneurs overtook government-ordered logging. For instance, when yet another fire broke out in Edo, one of the most famous of those private lumbermen, a merchant named Kinokuniya Bunzaemon, shrewdly recognized that the result would be more demand for timber. Even before the fire had been put out, he sailed off on a ship to buy up huge quantities of timber in the Kiso district, for resale at a big profit in Edo.

The first part of Japan to become deforested, already by A.D. 800, was the Kinai Basin on the largest

Japanese island of Honshu, site of early Japan's main cities such as Osaka and Kyoto. By the year 1000, deforestation was spreading to the nearby smaller island of Shikoku. By 1550 about one-quarter of Japan's area (still mainly just central Honshu and eastern Shikoku) had been logged, but other parts of Japan still held much lowland forest and old-growth forest.

In 1582 Hideyoshi became the first ruler to demand timber from all over Japan, because timber needs for his lavish monumental construction exceeded the timber available on his own domains. He took control of some of Japan's most valuable forests and requisitioned a specified amount of timber each year from each daimyo. In addition to forests, which the shogun and daimyo claimed for themselves, they also claimed all valuable species of timber trees on village or private land. To transport all that timber from increasingly distant logging areas to the cities or castles where the timber was needed, the government cleared obstacles from rivers so that logs could be floated or rafted down them to the coast, whence they were then transported by ships to port cities. Logging spread over Japan's three main islands, from the southern end of the southernmost island of Kyushu through Shikoku to the northern end of Honshu. In 1678 loggers had to turn to the southern end of Hokkaido, the island north of Honshu and at that time not yet part of the Japanese state. By 1710, most accessible forest had been cut on the three main islands (Kyushu, Shikoku, and Honshu) and on southern Hokkaido, leaving old-growth forests just on steep slopes, in inaccessible areas, and at sites too difficult or costly to log with Tokugawa-era technology.

Deforestation hurt Tokugawa Japan in other ways besides the obvious one of wood shortages for timber, fuel, and fodder and the forced end to monumental construction. Disputes over timber and fuel became increasingly frequent between and within villages, and between villages and the daimyo or shogun, all of whom competed for Japan's forests. There were also disputes between those who wanted to use rivers for floating or rafting logs, and those who instead wanted to use them for fishing or for irrigating cropland. Just as we saw for Montana in Chapter 1, wildfires increased, because the second-growth woods springing up on logged land were more flammable than were old-growth forests. Once the forest cover protecting steep slopes had been removed, the rate of soil erosion increased as a consequence of Japan's heavy rainfall, snowmelt, and frequent earthquakes. Flooding in the lowlands due to increased water runoff from the denuded slopes, higher water levels in lowland irrigation systems due to soil erosion and river siltation, increased storm damage, and shortages of forest-derived fertilizer and fodder acted together to decrease crop yields at a time of increasing population, and thus to contribute to major famines that beset Tokugawa Japan from the late 1600s onwards.

The 1657 Meireki fire, and the resulting demand for timber to rebuild Japan's capital, served as a wake-up call exposing the country's growing scarcity of timber and other resources at a time when its population, especially its urban population, had been growing rapidly. That might have led to an Easter Island–like catastrophe. Instead, over the course of the next two centuries Japan gradually achieved a stable population and much more nearly sustainable resource consumption rates. The shift was led from the top by successive shoguns, who invoked Confucian principles to promulgate an official ideology that encouraged limiting consumption and accumulating reserve supplies in order to protect the country against disaster.

Part of the shift involved increased reliance on seafood and on trade with the Ainu for food, in order to relieve the pressure on farming. Expanded fishing efforts incorporated new fishing techniques, such as very large nets and deepwater fishing. The territories claimed by individual daimyo and villages now included the sea adjacent to their land, in recognition of the sense that fish and shellfish stocks were limited and might become exhausted if anyone else could freely fish in one's territory. Pressure on forests as a source of green fertilizer for cropland was reduced by making much more use of fish meal fertilizers. Hunting of sea mammals (whales, seals, and sea otters) increased, and syndicates were formed to finance the necessary boats, equipment, and large workforces. The greatly expanded trade with the Ainu on Hokkaido Island brought smoked salmon, dried sea cucumber, abalone, kelp, deer skins, and sea otter pelts to Japan, in exchange for rice, sake (rice wine), tobacco, and cotton delivered to the Ainu. Among the results were the depletion of salmon and deer on Hokkaido, the weaning of the Ainu away from self-sufficiency as hunters to dependence on Japanese imports, and eventually the destruction of the Ainu through economic disruption, disease epidemics, and military conquests. Thus, part of the Tokugawa solution for the problem of resource depletion in Japan itself was to conserve Japanese resources by causing resource depletion elsewhere, just as part of the solution of Japan and other First World countries to problems of resource depletion today is to cause resource depletion elsewhere. (Remember that Hokkaido was not incorporated politically into Japan until the 19th century.)

Another part of the shift consisted of the near-achievement of Zero Population Growth. Between 1721 and 1828, Japan's population barely increased at all, from 26,100,000 to only 27,200,000. Compared to earlier centuries, Japanese in the 18th and 19th century married later, nursed their babies for longer, and spaced their children at longer intervals through the resulting lactational amenorrhea as well as through contraception, abortion, and infanticide. Those decreased birth rates represented responses of individual couples to perceived shortages of food and other resources, as shown by rises and falls in

Tokugawa Japanese birth rates in phase with falls and rises in rice prices.

Still other aspects of the shift served to reduce wood consumption. Beginning in the late 17th century, Japan's use of coal instead of wood as a fuel rose. Lighter construction replaced heavy-timbered houses, fuel-efficient cooking stoves replaced open-hearth fireplaces, small portable charcoal heaters replaced the practice of heating the whole house, and reliance on the sun to heat houses during the winter increased.

Many top-down measures were aimed at curing the imbalance between cutting trees and producing trees, initially mainly by negative measures (reducing the cutting), then increasingly by positive measures as well (producing more trees). One of the first signs of awareness at the top was a proclamation by the shogun in 1666, just nine years after the Meireki fire, warning of the dangers of erosion, stream siltation, and flooding caused by deforestation, and urging people to plant seedlings. Beginning in that same decade, Japan launched a nationwide effort at all levels of society to regulate use of its forest, and by 1700 an elaborate system of woodland management was in place. In the words of historian Conrad Totman, the system focused on "specifying who could do what, where, when, how, how much, and at what price." That is, the first phase of the Tokugawa-era response to Japan's forest problem emphasized negative measures that didn't restore lumber production to previous levels, but that at least bought time, prevented the situation from getting worse until positive measures could take effect, and set ground rules for the competition within Japanese society over increasingly scarce forest products.

The negative responses aimed at three stages in the wood supply chain: woodland management, wood transport, and wood consumption in towns. At the first stage, the shogun, who directly controlled about a quarter of Japan's forests, designated a senior magistrate in the finance ministry to be responsible for his forests, and almost all of the 250 daimyo followed suit by each appointing his own forest magistrate for his land. Those magistrates closed off logged lands to permit forest regeneration, issued licenses specifying the peasants' rights to cut timber or graze animals on government forest land, and banned the practice of burning forests to clear land for shifting cultivation. In those forests controlled not by the shogun or daimyo but by villages, the village headman managed the forest as common property for the use of all villagers, developed rules about the harvesting of forest products, forbade "foreign" peasants of other villages to use his own village's forest, and hired armed guards to enforce all these rules.

Both the shogun and the daimyo paid for very detailed inventories of their forests. Just as one example of the managers' obsessiveness, an inventory of a forest near Karuizawa 80 miles northwest of Edo in 1773 recorded that the forest measured 2.986 square miles in area and contained 4,114 trees, of which 573 were crooked or knotty and 3,541 were good. Of those 4,114 trees, 78 were big conifers (66 of them good) with trunks 24–36 feet long and 6–7 feet in circumference, 293 were medium-sized conifers (253 of them good) 4–5 feet in circumference, 255 good small conifers 6–18 feet long and 1–3 feet in circumference to be harvested in the year 1778, and 1,474 small conifers (1,344 of them good) to harvest in later years. There were also 120 medium-sized ridgeline conifers (104 of them good) 15–18 feet long and 3–4 feet in circumference, 15 small ridgeline conifers 12–24 feet long and 8 inches to 1 foot in circumference to be harvested in 1778, and 320 small ridgeline conifers (241 of them good) to harvest in later years, not to mention 448 oaks (412 of them good) 12–24 feet long and 3–5½ feet in circumference, and 1,126 other trees whose properties were similarly enumerated. Such counting represents an extreme of top-down management that left nothing to the judgment of individual peasants.

The second stage of negative responses involved the shogun and daimyo establishing guard posts on highways and rivers to inspect wood shipments and make sure that all those rules about woodland management were actually being obeyed. The last stage consisted of a host of government rules specifying, once a tree had been felled and had passed inspection at a guard post, who could use it for what purpose. Valuable cedars and oaks were reserved for government uses and were off limits to peasants. The amount of timber that you could use in building your house varied with your social status: 30 *ken* (one ken is a beam 6 feet long) for a headman presiding over several villages, 18 ken for such a headman's heir, 12 ken for a headman of a single village, 8 ken for a local chief, 6 ken for a taxable peasant, and a mere 4 ken for an ordinary peasant or fisherman. The shogun also issued rules about permissible wood use for objects smaller than houses. For instance, in 1663 an edict forbade any woodworker in Edo to fabricate a small box out of cypress or sugi wood, or household utensils out of sugi wood, but permitted large boxes to be made of either cypress or sugi. In 1668 the shogun went on to ban use of cypress, sugi, or any other good tree for public signboards, and 38 years later large pines were removed from the list of trees approved for making New Year decorations.

All of these negative measures aimed at solving Japan's forestry crisis by ensuring that wood be used only for purposes authorized by the shogun or daimyo. However, a big role in Japan's crisis had been played by wood use by the shogun and daimyo themselves. Hence a full solution to the crisis required positive measures to produce more trees, as well as to protect land from erosion. Those measures began already in the 1600s with Japan's development of a detailed body of scientific knowledge about silviculture. Foresters employed both by the government and by private merchants observed, experimented, and published their findings in an outpouring of silvi-cultural journals and manuals, exemplified by the first of Japan's great silvicultural treatises, the *Nōgyō zensho* of 1697 by Miyazaki Antei. There, you will find instructions for how best to gather, extract, dry, store, and prepare seeds; how to prepare a seedbed by cleaning, fertilizing, pulverizing, and stirring it; how to soak seeds before sowing them; how to protect sown seeds by spreading straw over them; how to weed the seedbed; how to transplant and space seedlings; how to replace failed seedlings over the next four years; how to thin out the resulting saplings; and how to trim branches from the growing trunk in order that it yield a log of the desired shape. As an alternative to thus growing trees from seed, some tree species were instead grown by planting cuttings or shoots, and others by the technique known as coppicing (leaving live stumps or roots in the ground to sprout).

Gradually, Japan independently of Germany developed the idea of plantation forestry: that trees should be viewed as a slow-growing crop. Both governments and private entrepreneurs began planting forests on land that they either bought or leased, especially in areas where it would be economically favorable, such as near cities where wood was in demand. On the one hand, plantation forestry is expensive, risky, and demanding of capital. There are big costs up front to pay workers to plant the trees, then more labor costs for several decades to tend the plantation, and no recovery of all that investment until the trees are big enough to harvest. At any time during those decades, one may lose one's tree crop to disease or a fire, and the price that the lumber will eventually fetch is subject to market fluctuations unpredictable decades in advance when the seeds are planted. On the other hand, plantation forestry offers several compensating advantages compared to cutting naturally sown forests. You can plant just preferred valuable tree species, instead of having to accept whatever sprouts in the forest. You can maximize the quality of your trees and the price received for them, for instance by trimming them as they grow to obtain eventually straight and well-shaped logs. You can pick a convenient site with low transport costs near a city and near a river suitable for floating logs out, instead of having to haul logs down a remote mountainside. You can space out your trees at equal intervals, thereby reducing the costs of eventual cutting. Some Japanese plantation foresters specialized in wood for particular uses and were thereby able to command top prices for an established "brand name." For instance, Yoshino plantations became known for producing the best staves for cedar barrels to hold sake (rice wine).

The rise of silviculture in Japan was facilitated by the fairly uniform institutions and methods over the whole country. Unlike the situation in Europe, divided at that time among hundreds of principalities or states, Tokugawa Japan was a single country governed uniformly. While southwestern Japan is subtropical and northern Japan is temperate, the whole country is alike in being wet, steep, erodable, of volcanic origins, and divided between steep forested mountains and flat cropland, thus providing some ecological uniformity in conditions for silviculture. In place of Japan's tradition of multiple use of forests, under which the elite claimed the timber and the peasants gathered fertilizer, fodder, and fuel, plantation forest became specified as being for the primary purpose of timber production, other uses being allowed only insofar as they did not harm timber production. Forest patrols guarded against illegal logging activity. Plantation forestry thereby became widespread in Japan between 1750 and 1800, and by 1800 Japan's long decline in timber production had been reversed.

An outside observer who visited Japan in 1650 might have predicted that Japanese society was on the verge of a societal collapse triggered by catastrophic deforestation, as more and more people competed for fewer resources. Why did Tokugawa Japan succeed in developing top-down solutions and thereby averting deforestation, while the ancient Easter Islanders, Maya, and Anasazi, and modern Rwanda (Chapter 10) and Haiti (Chapter 11) failed? This question is one example of the broader problem, to be explored in Chapter 14, why and at what stages people succeed or fail at group decision-making.

The usual answers advanced for Middle and Late Tokugawa Japan's success—a supposed love for Nature, Buddhist respect for life, or a Confucian outlook—can be quickly dismissed. In addition to those simple phrases not being accurate descriptions of the complex reality of Japanese attitudes, they did not prevent Early Tokugawa Japan from depleting Japan's resources, nor are they preventing modern Japan from depleting the resources of the ocean and of other countries today. Instead, part of the answer involves Japan's environmental advantages: some of the same environmental factors already discussed in Chapter 2 to explain why Easter and several other Polynesian and Melanesian islands ended up deforested, while Tikopia, Tonga, and others did not. People of the latter islands have the good fortune to be living in ecologically robust landscapes where trees regrow rapidly on logged soils. Like robust Polynesian and Melanesian islands, Japan has rapid tree regrowth because of high rainfall, high fallout of volcanic ash and Asian dust restoring soil fertility, and young soils. Another part of the answer has to do with Japan's social advantages: some features of Japanese society that already existed before the deforestation crisis and did not have to arise as a response to it. Those features included Japan's lack of goats and sheep, whose grazing and browsing activities elsewhere have devastated forests of many lands; the decline in number of horses in Early Tokugawa Japan, due to the end of warfare eliminating the need for cavalry; and the abundance of seafood, relieving pressure on forests as sources of protein and fertilizer. Japanese society did make use of oxen and horses as draft animals, but their numbers were allowed to decrease in response to deforestation and loss of forest fodder, to be replaced by people using spades, hoes, and other devices.

The remaining explanations constitute a suite of factors that caused both the elite and the masses in Japan to recognize their long-term stake in preserving their own forests, to a degree greater than for most other people. As for the elite, the Tokugawa shoguns, having imposed peace and eliminated rival armies at home, correctly anticipated that they were at little risk of a revolt at home or an invasion from overseas. They expected their own Tokugawa family to remain in control of Japan, which in fact it did for 250 years. Hence peace, political stability, and well-justified confidence in their own future encouraged Tokugawa shoguns to invest in and to plan for the long-term future of their domain: in contrast to Maya kings and to Haitian and Rwandan presidents, who could not or cannot expect to be succeeded by their sons or even to fill out their own term in office. Japanese society as a whole was (and still is) relatively homogeneous ethnically and religiously, without the differences destabilizing Rwandan society and possibly also Maya and Anasazi societies. Tokugawa Japan's isolated location, negligible foreign trade, and renunciation of foreign expansion made it obvious that it had to depend on its own resources and wouldn't solve its needs by pillaging another country's resources. By the same token, the shogun's enforcement of peace within Japan meant that people knew that they couldn't meet their timber needs by seizing a Japanese neighbor's timber. Living in a stable society without input of foreign ideas, Japan's elite and peasants alike expected the future to be like the present, and future problems to have to be solved with present resources.

The usual assumption of Tokugawa well-to-do peasants, and the hope of poorer villagers, were that their land would pass eventually to their own heirs. For that and other reasons, the real control of Japan's forests fell increasingly into the hands of people with a vested long-term interest in their forest: either because they thus expected or hoped their children would inherit the rights to its use, or because of various long-term lease or contract arrangements. For instance, much village common land became divided into separate leases for individual households, thereby minimizing the tragedies of the common to be discussed in Chapter 14. Other village forests were managed under timber sale agreements drawn up long in advance of logging. The government negotiated long-term contracts on government forest land, dividing eventual timber proceeds with a village or merchant in return for the latter managing the forests. All these political and social factors made it in the interests of the shogun, daimyo, and peasants to manage their forests sustainably. Equally obviously after the Meireki fire, those factors made short-term Overexploitation of forests foolish.

Of course, though, people with long-term stakes don't always act wisely. Often they still prefer short-term goals, and often again they do things that are foolish in both the short term and the long term. That's what makes biography and history infinitely more complicated and less predictable than the courses of chemical reactions, and that's why this book doesn't preach environmental determinism. Leaders who don't just react passively, who have the courage to anticipate crises or to act early, and who make strong insightful decisions of top-down management really can make a huge difference to their societies. So can similarly courageous, active citizens practicing bottom-up management. The Tokugawa shoguns, and my Montana landowner friends committed to the Teller Wildlife Refuge, exemplify the best of each type of management, in pursuit of their own long-term goals and of the interests of many others.

In thus devoting one chapter to these three success stories of the New Guinea highlands, Tikopia, and Tokugawa Japan, after seven chapters mostly on societies brought down by deforestation and other environmental problems plus a few other success stories (Orkney, Shetland, Faeroes, Iceland), I'm not implying that success stories constitute rare exceptions. Within the last few centuries Germany, Denmark, Switzerland, France, and other western European countries stabilized and then expanded their forested area by top-down measures, as did Japan. Similarly, about 600 years earlier, the largest and most tightly organized Native American society, the Inca Empire of the Central Andes with tens of millions of subjects under an absolute ruler, carried out massive reafforestation and terracing to halt soil erosion, increase crop yields, and secure its wood supplies.

Examples of successful bottom-up management of small-scale farming, pastoral, hunting, or fishing economies also abound. One example that I briefly mentioned in Chapter 4 comes from the U.S. Southwest, where Native American societies far smaller than the Inca Empire attempted many different solutions to the problem of developing a long-lasting economy in a difficult environment. The Anasazi, Hohokam, and Mimbres solutions eventually came to an end, but the somewhat different Pueblo solution has now been operating in the same region for over a thousand years. While the Greenland Norse disappeared, the Greenland Inuit maintained a self-sufficient hunter-gatherer economy for at least 500 years, from their arrival by A.D. 1200 until the disruptions caused by Danish colonization beginning in A.D. 1721. After the extinction of Australia's Pleistocene megafauna around 46,000 years ago, Aboriginal Australians maintained hunter-gatherer economies until European settlement in A.D. 1788. Among the numerous, self-sustaining, small-scale rural societies in modern times, especially well-studied ones include communities in Spain and in the Philippines maintaining irrigation systems, and Swiss alpine villages operating mixed farming and pastoral economies, in both cases for many centuries and with detailed local agreements about managing communal resources.

Each of these cases of bottom-up management that I have just mentioned involves a small society holding exclusive rights to all economic activities on its lands. Interesting and more complex cases exist (or traditionally existed) on the Indian subcontinent, where the caste system instead operates to permit dozens of economically specialized sub-societies to share the same geographic area by carrying out different economic activities. Castes trade extensively with each other and often live in the same village but are endogamous—i.e., people generally marry within their caste. Castes coexist by exploiting different environmental resources and lifestyles, such as by fishing, farming, herding, and hunting/gathering. There is even finer specialization, e.g., with multiple castes of fishermen fishing by different methods in different types of waters. As in the case of Tikopians and of the Tokugawa Japanese, members of the specialized Indian castes know that they can count on only a circumscribed resource base to maintain themselves, but they expect to pass those resources on to their children. Those conditions have fostered the acceptance of very detailed societal norms by which members of a given caste ensure that they are exploiting their resources sustainably.

The question remains why these societies of Chapter 9 succeeded while most of the societies selected for discussion in Chapters 2–8 failed. Part of the explanation lies in environmental differences: some environments are more fragile and pose more challenging problems than do others. We already saw in Chapter 2 the multitude of reasons causing Pacific island environments to be more or less fragile, and explaining in part why Easter and Mangareva societies collapsed while Tikopia society didn't. Similarly, the success stories of the New Guinea highlands and Tokugawa Japan recounted in this chapter involved societies that enjoyed the good fortune to be occupying relatively robust environments. But environmental differences aren't the whole explanation, as proved by the cases, such as those of Greenland and the U.S. Southwest, in which one society succeeded while one or more societies practicing different economies in the same environment failed. That is, not only the environment, but also the proper choice of an economy to fit the environment, is important. The remaining large piece of the puzzle involves whether, even for a particular type of economy, a society practices it sustainably. Regardless of the resources on which the economy rests—farmed soil, grazed or browsed vegetation, a fishery, hunted game, or gathered plants or small animals—some societies evolve practices to avoid overexploitation, and other societies fail at that challenge.

———————

[Diamond delineates twelve sustainability problems that face modern societies:

1. Habitat destruction
2. Destruction of wild food supply, particularly fisheries
3. Biodiversity destruction, which damages ecosystem functioning
4. Soil erosion and soil damage
5. Limits to the fossil fuel supply
6. Freshwater shortages
7. Overuse of planetary photosynthetic capacity for food, wood, biofuels, etc.
8. Toxic chemicals like pesticides, heavy metals, and components of plastics such as BPA and other endocrine disruptors
9. Invasive species that disrupt ecosystems and infect humans
10. Trace gas effects including global warming and destruction of the ozone layer
11. Overpopulation (too many people)
12. Over-affluence (too much stuff/environmental impact per person)

Of these 12 problems, 1) thru 4) deal with natural resource loss, 5)–7) with limits on natural resources, 8)–10) with pollution and 11)–12) with population. 5, 7, 8, and 10 are modern problems; the rest have been around for a long time. Here is a detailed description, as concise a delineation of the planetary resources crisis as I have encountered:]

―――――――――

1. At an accelerating rate, we are destroying natural habitats or else converting them to human-made habitats, such as cities and villages, farmlands and pastures, roads, and golf courses. The natural habitats whose losses have provoked the most discussion are forests, wetlands, coral reefs, and the ocean bottom. As I mentioned in the preceding chapter, more than half of the world's original area of forest has already been converted to other uses, and at present conversion rates one-quarter of the forests that remain will become converted within the next half-century. Those losses of forests represent losses for us humans, especially because forests provide us with timber and other raw materials, and because they provide us with so-called ecosystem services such as protecting our watersheds, protecting soil against erosion, constituting essential steps in the water cycle that generates much of our rainfall, and providing habitat for most terrestrial plant and animal species. Deforestation was a or *the* major factor in all the collapses of past societies described in this book. In addition, as discussed in Chapter 1 in connection with Montana, issues of concern to us are not only forest destruction and conversion, but also changes in the structure of wooded habitats that do remain. Among other things, that changed structure results in changed fire regimes that put forests, chaparral woodlands, and savannahs at greater risk of infrequent but catastrophic fires.

Other valuable natural habitats besides forests are also being destroyed. An even larger fraction of the world's original wetlands than of its forests has already been destroyed, damaged, or converted. Consequences for us arise from wetlands' importance in maintaining the quality of our water supplies and the existence of commercially important freshwater fisheries, while even ocean fisheries depend on mangrove wetlands to provide habitat for the juvenile phase of many fish species. About one-third of the world's coral reefs—the oceanic equivalent of tropical rainforests, because they are home to a disproportionate fraction of the ocean's species—have already been severely damaged. If current trends continue, about half of the remaining reefs would be lost by the year 2030. That damage and destruction result from the growing use of dynamite as a fishing method, reef overgrowth by algae ("seaweeds") when the large herbivorous fish that normally graze on the algae become fished out, effects of sediment runoff and pollutants from adjacent lands cleared or converted to agriculture, and coral bleaching due to rising ocean water temperatures. It has recently become appreciated that fishing by trawling is destroying much or most of the shallow ocean bottom and the species dependent on it.

2. Wild foods, especially fish and to a lesser extent shellfish, contribute a large fraction of the protein consumed by humans. In effect, this is protein that we obtain for free (other than the cost of catching and transporting the fish), and that reduces our needs for animal protein that we have to grow ourselves in the form of domestic livestock. About two billion people, most of them poor, depend on the oceans for protein. If wild fish stocks were managed appropriately, the stock levels could be maintained, and they could be harvested perpetually. Unfortunately, the problem known as the tragedy of the commons (Chapter 14) has regularly undone efforts to manage fisheries sustainably, and the great majority of valuable fisheries already either have collapsed or are in steep decline (Chapter 15).

Past societies that overfished included Easter Island, Mangareva, and Henderson.

Increasingly, fish and shrimp are being grown by aquaculture, which in principle has a promising future as the cheapest way to produce animal protein. In several respects, though, aquaculture as commonly practiced today is making the problem of declining wild fisheries worse rather than better. Fish grown by aquaculture are mostly fed wild-caught fish and thereby usually consume more wild fish meat (up to 20 times more) than they yield in meat of their own They contain higher toxin levels than do wild-caught fish. Cultured fish regularly escape, interbreed with wild fish, and thereby harm wild fish stocks genetically, because cultured fish strains have been selected for rapid growth at the expense of poor survival in the wild (50 times worse survival for cultured salmon than for wild salmon). Aquaculture runoff causes pollution and eutrophication. The lower costs of aquaculture than of fishing, by driving down fish prices, initially drive fishermen to exploit wild fish stocks even more heavily in order to maintain their incomes constant when they are receiving less money per pound of fish.

3. A significant fraction of wild species, populations, and genetic diversity has already been lost, and at present rates a large fraction of what remains will be lost within the next half-century. Some species, such as big edible animals, or plants with edible fruits or good timber, are of obvious value to us. Among the many past societies that harmed themselves by exterminating such species were the Easter and Henderson Islanders whom we have discussed.

But biodiversity losses of small inedible species often provoke the response, "Who cares? Do you really care less for humans than for some lousy useless little fish or weed, like the snail darter or Furbish lousewort?" This response misses the point that the entire natural world is made up of wild species providing us for free with services that can be very expensive, and in many cases impossible, for us to supply ourselves. Elimination of lots of lousy little species regularly causes big harmful consequences for humans, just as does randomly knocking out many of the lousy little rivets holding together an airplane. The literally innumerable examples include: the role of earthworms in regenerating soil and maintaining its texture (one of the reasons that oxygen levels dropped inside the Biosphere 2 enclosure, harming its human inhabitants and crippling a colleague of mine, was a lack of appropriate earthworms, contributing to altered soil/atmosphere gas exchange); soil bacteria that fix the essential crop nutrient nitrogen, which otherwise we have to spend money to supply in fertilizers; bees and other insect pollinators (they pollinate our crops for free, whereas it's expensive for us to pollinate every crop flower by hand); birds and mammals that disperse wild fruits (foresters still haven't figured out how to grow from seed the most important commercial tree species of the Solomon Islands, whose seeds are naturally dispersed by fruit bats, which are becoming hunted out); elimination of whales, sharks, bears, wolves, and other top predators in the seas and on the land, changing the whole food chain beneath them; and wild plants and animals that decompose wastes and recycle nutrients, ultimately providing us with clean water and air.

4. Soils of farmlands used for growing crops are being carried away by water and wind erosion at rates between 10 and 40 times the rates of soil formation, and between 500 and 10,000 times soil erosion rates on forested land. Because those soil erosion rates are so much higher than soil formation rates, that means a net loss of soil. For instance, about half of the top-soil of Iowa, the state whose agriculture productivity is among the highest in the U.S., has been eroded in the last 150 years. On my most recent visit to Iowa, my hosts showed me a churchyard offering a dramatically visible example of those soil losses. A church was built there in the

middle of farmland during the 19th century and has been maintained continuously as a church ever since, while the land around it was being farmed. As a result of soil being eroded much more rapidly from fields than from the churchyard, the yard now stands like a little island raised 10 feet above the surrounding sea of farmland.

Other types of soil damage caused by human agricultural practices include salinization, as discussed for Montana, China, and Australia in Chapters 1, 12, and 13; losses of soil fertility, because farming removes nutrients much more rapidly than they are restored by weathering of the underlying rock; and soil acidification in some areas, or its converse, alkalinization, in other areas. All of these types of harmful impacts have resulted in a fraction of the world's farmland variously estimated at between 20% and 80% having become severely damaged, during an era in which increasing human population has caused us to need more farmland rather than less farmland. Like deforestation, soil problems contributed to the collapses of all past societies discussed in this book.

The next three problems involve ceilings—on energy, freshwater, and photosynthetic capacity. In each case the ceiling is not hard and fixed but soft: we can obtain more of the needed resource, but at increasing costs.

5. The world's major energy sources, especially for industrial societies, are fossil fuels: oil, natural gas, and coal. While there has been much discussion about how many big oil and gas fields remain to be discovered, and while coal reserves are believed to be large, the prevalent view is that known and likely reserves of readily accessible oil and natural gas will last for a few more decades. This view should not be misinterpreted to mean that all of the oil and natural gas within the Earth will have been used up

by then. Instead, further reserves will be deeper underground, dirtier, increasingly expensive to extract or process, or will involve higher environmental costs. Of course, fossil fuels are not our sole energy sources, and I shall consider problems raised by the alternatives below.

6. Most of the world's freshwater in rivers and lakes is already being utilized for irrigation, domestic and industrial water, and in situ uses such as boat transportation corridors, fisheries, and recreation. Rivers and lakes that are not already utilized are mostly far from major population centers and likely users, such as in Northwestern Australia, Siberia, and Iceland. Throughout the world, freshwater underground aquifers are being depleted at rates faster than they are being naturally replenished, so that they will eventually dwindle. Of course, freshwater can be made by desalinization of seawater, but that costs money and energy, as does pumping the resulting desalinized water inland for use. Hence desalinization, while it is useful locally, is too expensive to solve most of the world's water shortages. The Anasazi and Maya were among the past societies to be undone by water problems, while today over a billion people lack access to reliable safe drinking water.

7. It might at first seem that the supply of sunlight is infinite, so one might reason that the Earth's capacity to grow crops and wild plants is also infinite. Within the last 20 years, it has been appreciated that that is not the case, and that's not only because plants grow poorly in the world's Arctic regions and deserts unless one goes to the expense of supplying heat or water. More generally, the amount of solar energy fixed per acre by plant photosynthesis, hence plant growth per acre, depends on temperature and rainfall. At any given temperature and rainfall the plant growth that can be supported by the sunlight falling on an acre is limited by the geometry and biochemistry of plants, even if they take up the sunlight so efficiently that not a single photon of light

passes through the plants unabsorbed to reach the ground. The first calculation of this photosynthetic ceiling, carried out in 1986, estimated that humans then already used (e.g., for crops, tree plantations, and golf courses) or diverted or wasted (e.g., light falling on concrete roads and buildings) about half of the Earth's photosynthetic capacity. Given the rate of increase of human population, and especially of population impact (see point 12 below), since 1986, we are projected to be utilizing most of the world's terrestrial photosynthetic capacity by the middle of this century. That is, most energy fixed from sunlight will be used for human purposes, and little will be left over to support the growth of natural plant communities, such as natural forests.

The next three problems involve harmful things that we generate or move around: toxic chemicals, alien species, and atmospheric gases.

8. The chemical industry and many other industries manufacture or release into the air, soil, oceans, lakes, and rivers many toxic chemicals, some of them "unnatural" and synthesized only by humans, others present naturally in tiny concentrations (e.g., mercury) or else synthesized by living things but synthesized and released by humans in quantities much larger than natural ones (e.g., hormones). The first of these toxic chemicals to achieve wide notice were insecticides, pesticides, and herbicides, whose effects on birds, fish, and other animals were publicized by Rachel Carson's 1962 book *Silent Spring*. Since then, it has been appreciated that the toxic effects of even greater significance for us humans are those on ourselves. The culprits include not only insecticides, pesticides, and herbicides, but also mercury and other metals, fire-retardant chemicals, refrigerator coolants, detergents, and components of plastics. We swallow them in our food and water, breathe them in our air, and absorb them through our skin. Often in very low concentrations, they variously cause birth defects, mental retardation, and temporary or permanent damage to our immune and reproductive systems. Some of them act as endocrine disruptors, i.e., they interfere with our reproductive systems by mimicking or blocking effects of our own sex hormones. They probably make the major contribution to the steep decline in sperm count in many human populations over the last several decades, and to the apparently increasing frequency with which couples are unable to conceive, even when one takes into account the increasing average age of marriage in many societies. In addition, deaths in the U.S. from air pollution alone (without considering soil and water pollution) are conservatively estimated at over 130,000 per year.

Many of these toxic chemicals are broken down in the environment only slowly (e.g., DDT and PCBs) or not at all (mercury), and they persist in the environment for long times before being washed out. Thus, cleanup costs of many polluted sites in the U.S. are measured in the billions of dollars (e.g., Love Canal, the Hudson River, Chesapeake Bay, the *Exxon Valdez* oil spill, and Montana copper mines). But pollution at those worst sites in the U.S. is mild compared to that in the former Soviet Union, China, and many Third World mines, whose cleanup costs no one even dares to think about.

9. The term "alien species" refers to species that we transfer, intentionally or inadvertently, from a place where they are native to another place where they are not native. Some alien species are obviously valuable to us as crops, domestic animals, and landscaping. But others devastate populations of native species with which they come in contact, either by preying on, parasitizing, infecting, or outcompeting them. The aliens cause these big effects because the native species with which they come in contact had no previous evolutionary experience of them and are unable to resist them (like human populations

newly exposed to smallpox or AIDS). There are by now literally hundreds of cases in which alien species have caused one-time or annually recurring damages of hundreds of millions of dollars or even billions of dollars. Modern examples include Australia's rabbits and foxes, agricultural weeds like Spotted Knapweed and Leafy Spurge (Chapter 1), pests and pathogens of trees and crops and livestock (like the blights that wiped out American chestnut trees and devasted American elms), the water hyacinth that chokes waterways, the zebra mussels that choke power plants, and the lampreys that devastated the former commercial fisheries of the North American Great Lakes (Plates 30, 31). Ancient examples include the introduced rats that contributed to the extinction of Easter Island's palm tree by gnawing its nuts, and that ate the eggs and chicks of nesting birds on Easter, Henderson, and all other Pacific islands previously without rats.

10. Human activities produce gases that escape into the atmosphere, where they either damage the protective ozone layer (as do formerly widespread refrigerator coolants) or else act as greenhouse gases that absorb sunlight and thereby lead to global warming. The gases contributing to global warming include carbon dioxide from combustion and respiration, and methane from fermentation in the intestines of ruminant animals. Of course, there have always been natural fires and animal respiration producing carbon dioxide, and wild ruminant animals producing methane, but our burning of firewood and of fossil fuels has greatly increased the former, and our herds of cattle and of sheep have greatly increased the latter.

For many years, scientists debated the reality, cause, and extent of global warming: are world temperatures really historically high now, and, if so, by how much, and are humans the leading cause? Most knowledgeable scientists now agree that, despite year-to-year ups and downs of temperature that necessitate complicated analyses to extract warming trends, the atmosphere really has been undergoing an unusually rapid rise in temperature recently, and that human activities are the or a major cause. The remaining uncertainties mainly concern the future expected magnitude of the effect: e.g., whether average global temperatures will increase by "just" 1.5 degrees Centigrade or by 5 degrees Centigrade over the next century. Those numbers may not sound like a big deal, until one reflects that average global temperatures were "only" 5 degrees cooler at the height of the last Ice Age.

While one might at first think that we should welcome global warming on the grounds that warmer temperatures mean faster plant growth, it turns out that global warming will produce both winners and losers. Crop yields in cool areas with temperatures marginal for agriculture may indeed increase, while crop yields in already warm or dry areas may decrease. In Montana, California, and many other dry climates, the disappearance of mountain snowpacks will decrease the water available for domestic uses, and for irrigation that actually limits crop yields in those areas. The rise in global sea levels as a result of snow and ice melting poses dangers of flooding and coastal erosion for densely populated low-lying coastal plains and river deltas already barely above or even below sea level. The areas thereby threatened include much of the Netherlands, Bangladesh, and the seaboard of the eastern U.S., many low-lying Pacific islands, the deltas of the Nile and Mekong Rivers, and coastal and riverbank cities of the United Kingdom (e.g., London), India, Japan, and the Philippines. Global warming will also produce big secondary effects that are difficult to predict exactly in advance and that are likely to cause huge problems, such as further climate changes resulting from changes in ocean circulation resulting in turn from melting of the Arctic ice cap.

The remaining two problems involve the increase in human population:

11. The world's human population is growing. More people require more food, space, water, energy, and other resources. Rates and even the direction of human population change vary greatly around the world, with the highest rates of population growth (4% per year or higher) in some Third World countries, low rates of growth (1% per year or less) in some First World countries such as Italy and Japan, and negative rates of growth (i.e., decreasing populations) in countries facing major public health crises, such as Russia and AIDS-affected African countries. Everybody agrees that the world population is increasing, but that its annual percentage rate of increase is not as high as it was a decade or two ago. However, there is still disagreement about whether the world's population will stabilize at some value above its present level (double the present population?), and (if so) how many years (30 years? 50 years?) it will take for population to reach that level, or whether population will continue to grow.

There is long built-in momentum to human population growth because of what is termed the "demographic bulge" or "population momentum," i.e., a disproportionate number of children and young reproductive-age people in today's population, as a result of recent population growth. That is, suppose that every couple in the world decided tonight to limit themselves to two children, approximately the correct number of children to yield an unchanging population in the long run by exactly replacing their two parents who will eventually die (actually, around 2.1 children when one considers mortality, childless couples, and children who won't marry). The world's population would nevertheless continue to increase for about 70 years, because more people today are of reproductive age or entering reproductive age than are old and post-reproductive. The problem of human population growth has received much attention in recent decades and has given rise to movements such as Zero Population Growth, which aim to slow or halt the increase in the world's population.

12. What really counts is not the number of people alone, but their impact on the environment. If most of the world's 6 billion people today were in cryogenic storage and neither eating, breathing, nor metabolizing, that large population would cause no environmental problems. Instead, our numbers pose problems insofar as we consume resources and generate wastes. That per-capita impact—the resources consumed, and the wastes put out, by each person—varies greatly around the world, being highest in the First World and lowest in the Third World. On the average, each citizen of the U.S., western Europe, and Japan consumes 32 times more resources such as fossil fuels, and puts out 32 times more wastes, than do inhabitants of the Third World (Plate 35).

But low-impact people are becoming high-impact people for two reasons: rises in living standards in Third World countries whose inhabitants see and covet First World lifestyles; and immigration, both legal and illegal, of individual Third World inhabitants into the First World, driven by political, economic, and social problems at home. Immigration from low-impact countries is now the main contributor to the increasing populations of the U.S. and Europe. By the same token, the overwhelmingly most important human population problem for the world as a whole is not the high rate of population increase in Kenya, Rwanda, and some other poor Third World countries, although that certainly does pose a problem for Kenya and Rwanda themselves, and although that is the population problem most discussed. Instead, the biggest problem is the increase in total human impact, as the result of rising Third World living standards, and of Third World individuals moving to the First World and adopting First World living standards.

There are many "optimists" who argue that the world could support double its human population, and who consider only the increase in human numbers and not the average increase in per-capita impact. But I have not met anyone who seriously argues that the world could support 12 times its current impact, although an increase of that factor would result from all Third World inhabitants adopting First World living standards. (That factor of 12 is less than the factor of 32 that I mentioned in the preceding paragraph, because there are already First World inhabitants with high-impact lifestyles, although they are greatly outnumbered by Third World inhabitants.) Even if the people of China alone achieved a First World living standard while everyone else's living standard remained constant, that would double our human impact on the world (Chapter 12).

People in the Third World aspire to First World living standards. They develop that aspiration through watching television, seeing advertisements for First World consumer products sold in their countries, and observing First World visitors to their countries. Even in the most remote villages and refugee camps today, people know about the outside world. Third World citizens are encouraged in that aspiration by First World and United Nations development agencies, which hold out to them the prospect of achieving their dream if they will only adopt the right policies, like balancing their national budgets, investing in education and infrastructure, and so on.

But no one in First World governments is willing to acknowledge the dream's impossibility: the unsustainability of a world in which the Third World's large population were to reach and maintain current First World living standards. It is impossible for the First World to resolve that dilemma by blocking the Third World's efforts to catch up: South Korea, Malaysia, Singapore, Hong Kong, Taiwan, and Mauritius have already succeeded or are close to success; China and India are progressing rapidly by their own efforts; and the 15 rich Western European countries making up the European Union have just extended Union membership to 10 poorer countries of Eastern Europe, in effect thereby pledging to help those 10 countries catch up. Even if the human populations of the Third World did not exist, it would be impossible for the First World alone to maintain its present course, because it is not in a steady state but is depleting its own resources as well as those imported from the Third World. At present, it is untenable politically for First World leaders to propose to their own citizens that they lower their living standards, as measured by lower resource consumption and waste production rates. What will happen when it finally dawns on all those people in the Third World that current First World standards are unreachable for them, and that the First World refuses to abandon those standards for itself? Life is full of agonizing choices based on trade-offs, but that's the cruelest trade-off that we shall have to resolve: encouraging and helping all people to achieve a higher standard of living, without thereby undermining that standard through overstressing global resources.

I introduced this section by acknowledging that there are important differences between the ancient world and the modern world. The differences that I then went on to mention—today's larger population and more potent destructive technology, and today's interconnectedness posing the risk of a global rather than a local collapse—may seem to suggest a pessimistic outlook. If the Easter Islanders couldn't solve their milder local problems in the past, how can the modern world hope to solve its big global problems?

People who get depressed at such thoughts often then ask me, "Jared, are you optimistic or pessimistic about the world's future?" I answer, "I'm a cautious optimist." By that, I mean that, on the one hand, I acknowledge the seriousness of the problems facing us. If we don't make a determined effort to solve them, and if we don't succeed at that effort, the world as a whole within the next few decades will face a declining standard of living, or perhaps something worse. That's the reason why I decided to devote most of my career efforts at this stage of my life to convincing people that our problems have to be taken seriously and won't go away otherwise. On the other hand, we shall be able to solve our problems—if we choose to do so. That's why my wife and I did decide to have children 17 years ago: because we did see grounds for hope.

One basis for hope is that, realistically, we are not beset by insoluble problems. While we do face big risks, the most serious ones are not ones beyond our control, like a possible collision with an asteroid of a size that hits the Earth every hundred million years or so. Instead, they are ones that we are generating ourselves. Because we are the cause of our environmental problems, we are the ones in control of them, and we can choose or not choose to stop causing them and start solving them. The future is up for grabs, lying in our own hands. We don't need new technologies to solve our problems; while new technologies can make some contribution, for the most part we "just" need the political will to apply solutions already available. Of course, that's a big "just." But many societies did find the necessary political will in the past. Our modern societies have already found the will to solve some of our problems, and to achieve partial solutions to others.

Another basis for hope is the increasing diffusion of environmental thinking among the public around the world. While such thinking has been with us for a long time, its spread has accelerated, especially since the 1962 publication of *Silent Spring*. The environmental movement has been gaining adherents at an increasing rate, and they act through a growing diversity of increasingly effective organizations, not only in the United States and Europe but also in the Dominican Republic and other developing countries. At the same time as the environmental movement is gaining strength at an increasing rate, so too are the threats to our environment. That's why I referred earlier in this book to our situation as that of being in an exponentially accelerating horse race of unknown outcome. It's neither impossible, nor is it assured, that our preferred horse will win the race.

What are the choices that we must make if we are now to succeed, and not to fail? There are many specific choices, of which I discuss examples in the Further Readings section, that any of us can make as individuals. For our society as a whole, the past societies that we have examined in this book suggest broader lessons. Two types of choices seem to me to have been crucial in tipping their outcomes towards success or failure: long-term planning, and willingness to reconsider core values. On reflection, we can also recognize the crucial role of these same two choices for the outcomes of our individual lives.

One of those choices has depended on the courage to practice long-term thinking, and to make bold, courageous, anticipatory decisions at a time when problems have become perceptible but before they have reached crisis proportions. This type of decision-making is the opposite of the short-term reactive decision-making that too often characterizes our elected politicians—the thinking that my politically well-connected friend decried as "90-day thinking," i.e., focusing only on issues likely to blow up in a crisis within the next 90 days. Set against the many depressing bad examples of such short-term decision-making are the encouraging examples of courageous long-term thinking in the past, and in the contemporary world of NGOs, business, and government. Among past societies faced with the prospect of ruinous deforestation, Easter Island and Mangareva chiefs succumbed to their immediate concerns, but Tokugawa shoguns, Inca emperors, New Guinea highlanders, and 16th-century German landowners adopted a long view and reafforested. China's leaders similarly promoted reafforestation in recent decades and banned logging of native forests in 1998. Today, many NGOs exist specifically for the purpose of promoting sane long-term environmental policies. In the business world the American corporations that remain successful for long times (e.g., Procter and Gamble) are ones that don't wait for a crisis to force them to reexamine their policies, but that instead look for problems on the horizon and act before there is a crisis. I already mentioned Royal Dutch Shell Oil Company as having an office devoted just to envisioning scenarios decades off in the future.

Courageous, successful, long-term planning also characterizes some governments and some political leaders, some of the time. Over the last 30 years a sustained effort by the U.S. government has reduced levels of the six major air pollutants nationally by 25%, even though our energy consumption and population increased by 40% and our vehicle miles driven increased by 150% during those same decades. The governments of Malaysia, Singapore, Taiwan, and Mauritius all recognized that their long-term economic well-being required big investments in public health to prevent tropical diseases from sapping their economies; those investments proved to be a key to those countries' spectacular recent economic growth. Of the former two halves of the overpopulated nation of Pakistan, the eastern half (independent since 1971 as Bangladesh) adopted effective family planning measures to reduce its rate of population growth, while the western half (still known as Pakistan) did not and is now the world's sixth most populous country. Indonesia's former environmental minister Emil Salim, and the Dominican Republic's former president Joaquín Balaguer, exemplify government leaders whose concern about chronic environmental dangers made a big impact on their countries. All of these examples of courageous long-term thinking in both the public sector and the private sector contribute to my hope.

The other crucial choice illuminated by the past involves the courage to make painful decisions about values. Which of the values that formerly served a society well can continue to be maintained under new changed circumstances? Which of those treasured values must instead be jettisoned and replaced with different approaches? The Greenland Norse refused to jettison part of their identity as a European, Christian, pastoral society, and they died as a result. In contrast, Tikopia Islanders did have the courage to eliminate their ecologically destructive pigs, even though pigs are the sole large domestic animal and a principal status symbol of Melanesian societies. Australia is now in the process of reappraising its identity as a British agricultural society. The Icelanders and many traditional caste societies of India in the past, and Montana ranchers dependent on irrigation in recent times, did reach agreement to subordinate their individual rights to group interests. They thereby succeeded in managing shared resources and avoiding the tragedy of the commons that has befallen so many other groups. The government of China restricted the traditional freedom of individual reproductive choice, rather than let population problems spiral out of control. The people of Finland, faced with an ultimatum by their vastly more powerful Russian neighbor in 1939, chose to value their freedom over their lives, fought with a courage that astonished the world, and won their gamble, even while losing the war. While I was living in Britain from 1958 to 1962, the British people were coming to terms with the outdatedness of cherished long-held values based on Britain's former role as the world's dominant political, economic, and naval power. The French, Germans, and other European countries have advanced even further in subordinating to the European Union their national sovereignties for which they used to fight so dearly.

All of these past and recent reappraisals of values that I have just mentioned were achieved despite being agonizingly difficult. Hence they also contribute to my hope. They may inspire modern First

World citizens with the courage to make the most fundamental reappraisal now facing us: how much of our traditional consumer values and First World living standard can we afford to retain? I already mentioned the seeming political impossibility of inducing First World citizens to lower their impact on the world. But the alternative, of continuing our current impact, is more impossible. This dilemma reminds me of Winston Churchill's response to criticisms of democracy: "It has been said that Democracy is the worst form of government except all those other forms that have been tried from time to time." In that spirit, a lower-impact society is the most impossible scenario for our future—except for all other conceivable scenarios.

Actually, while it won't be easier to reduce our impact, it won't be impossible either. Remember that impact is the product of two factors: population, multiplied times impact per person. As for the first of those two factors, population growth has recently declined drastically in all First World countries, and in many Third World countries as well—including China, Indonesia, and Bangladesh, with the world's largest, fourth largest, and ninth largest populations respectively. Intrinsic population growth in Japan and Italy is already below the replacement rate, such that their existing populations (i.e., not counting immigrants) will soon begin shrinking. As for impact per person, the world would not even have to decrease its current consumption rates of timber products or of seafood: those rates could be sustained or even increased, if the world's forests and fisheries were properly managed.

My remaining cause for hope is another consequence of the globalized modern world's interconnectedness. Past societies lacked archaeologists and television. While the Easter Islanders were busy deforesting the highlands of their overpopulated island for agricultural plantations in the 1400s, they had no way of knowing that, thousands of miles to the east and west at the same time, Greenland Norse society and the Khmer Empire were simultaneously in terminal decline, while the Anasazi had collapsed a few centuries earlier, Classic Maya society a few more centuries before that, and Mycenean Greece 2,000 years before that. Today, though, we turn on our television sets or radios or pick up our newspapers, and we see, hear, or read about what happened in Somalia or Afghanistan a few hours earlier. Our television documentaries and books show us in graphic detail why the Easter Islanders, Classic Maya, and other past societies collapsed. Thus, we have the opportunity to learn from the mistakes of distant peoples and past peoples. That's an opportunity that no past society enjoyed to such a degree. My hope in writing this book has been that enough people will choose to profit from that opportunity to make a difference.

When the Rivers Run Dry...

By Fred Pearce

Introduction

Oceans cover 70% of our planet, but these waters are salty. Usable freshwater is increasingly in short supply, and water supply is a potent source of international tension. In *When the Rivers Run Dry*, Fred Pearce covers many facets of the global water crisis, including the virtual water trade, the "mirage" of unsustainable groundwater pumping, the importance and fragility of our remaining wetlands, the incredible fecundity of undammed rivers, the catastrophic effects of dams, the role of climate change in drying the dry regions and soaking the wet ones, the ongoing role of river salts in killing productive farmland and even in destroying civilizations, and the powerful role of rivers in fueling international conflict. Long before global warming's effects become truly catastrophic, freshwater shortages will greatly challenge the international community. Most major rivers run through more than one country, giving power to the upstream parties over the downstream parties; when those parties are in conflict, there is high potential for severe conflict (even *nuclear* conflict, in the case of the Indian/Pakistani fight over the Indus River).

Pearce also gives us enormous hope. The concluding chapters of our selection cover sustainable responses to the freshwater crisis, including affordable drip irrigation in the developing world; the return to ancient methodologies such as rainwater harvesting and the *qanat* system of "wells" that never run dry; the safe recovery of nutrients and water from sewage; fog harvesting; and the total rethinking of urban stormwater management from running the water into the ocean as fast as possible, to holding onto and using valuable runoff. We humans are highly adaptable and capable of learning from our mistakes; let us pray we do so with this most precious of resources.

The Human Sponge

Few of us realize how much water it takes to get us through the day. On average, we drink no more than a gallon and a half of the stuff. Including water for washing and for flushing the toilet, we use only about 40 gallons each. In some countries suburban lawn sprinklers, swimming pools, and sundry outdoor uses can double that figure. Typical per capita water use in suburban Australia is about 90 gallons, and in the United States around 100 gallons. There are exceptions, though. One suburban household in Orange County, Florida, was billed for 4.1 million gallons in a single year, or more than 10,400 gallons a day. Nobody knows how they got through that much.

We can all save water in the home. But as laudable as it is to take a shower rather than a bath and turn off the faucet while brushing our teeth, we shouldn't get hold of the idea that regular domestic water use is what is really emptying the world's rivers. Manufacturing the goods that we fill our homes with consumes a certain amount, but that's not the real story either. It is only when we add in the water needed to grow what we eat and drink that the numbers really begin to soar.

Get your head around a few of these numbers, if you can. They are mind-boggling. It takes between 250 and 650 gallons of water to grow a pound of rice. That is more water than many households use in a week. For just a bag of rice. Keep going. It takes 130 gallons to grow a pound of wheat and 65 gallons for a pound of potatoes. And when you start feeding grain to livestock for animal products such as meat and milk, the numbers become yet more startling. It takes 3,000 gallons to grow the feed for enough cow to make a quarter-pound hamburger, and between 500 and 1,000 gallons for that cow to fill its udders with a quart of milk. Cheese? That takes about 650 gallons for a pound of cheddar or brie or camembert.

And if you think your shopping cart is getting a little bulky at this point, maybe you should leave that 1-pound box of sugar on the shelf. It took up to 400 gallons to produce. And the l-pound jar of coffee tips the scales at 2,650 gallons—or 10 tons—of water. Imagine taking *that* home from the store.

Turn these statistics into meal portions and you come up with more than 25 gallons for a portion of rice, 40 gallons for the bread in a sandwich or a serving of toast, 130 gallons for a two-egg omelet or a mixed salad, 265 gallons for a glass of milk, 400 gallons for an ice cream, 530 gallons for a pork chop, 800 gallons for a hamburger, and 1,320 gallons for a small steak. And if you have a sweet tooth, so much the worse: every teaspoonful of sugar in your coffee requires 50 cups of water to grow. Which is a lot, but not as much as the 37 gallons of water (or 592 cups) needed to grow the coffee itself. Prefer alcohol? A glass of wine or beer with dinner requires another 66 gallons, and a glass of brandy afterward takes a staggering 530 gallons.

We are all used to reading detailed technical information about the nutritional content of most food. Maybe it is time that we were given some clues as to how much water it took to grow and process the food. As the world's rivers run dry, it matters.

I figure that as a typical meat-eating, beer-swilling, milk-guzzling Westerner, I consume as much as a hundred times my own weight in water every day. Hats off, then, to my vegetarian daughter, who gets by with about half that. It's time, surely, to go out and preach the gospel of water conservation.

But don't buy one of those jokey T-shirts advertised on the Internet with slogans like "Save water, bathe with a friend." Good message, but you could fill roughly twenty-five bathtubs with the water needed to grow the 9 ounces of cotton needed to make the shirt. It gives a whole new meaning to the wet T-shirt contest.

Let's do the annual audit. I probably drink only about 265 gallons of water—that's one ton or 1.3 cubic yards—in a whole year. Around the home I probably use between 50 and 100 tons. But growing the crops to feed and clothe me for a year must take between 1,500 and 2,000 tons—more than half the contents of an Olympic-size swimming pool.

Where does all that water come from? In England, where I live, most homegrown crops are watered by rain. So the water is at least cheap. But remember that a lot of the food consumed in Britain, and all the cotton, is imported. And when the water to grow crops is collected from rivers or pumped from underground, as it is in much of the world, it is increasingly expensive, and its diversion to fields is increasingly likely to deprive someone else of water and to empty rivers and underground water reserves. And when the rivers are running low, it is ever more likely that the water simply will not be there to grow the crops at all.

The water "footprint" of Western countries on the rest of the world deserves to become a serious issue. Whenever you buy a T-shirt made of Pakistani cotton, eat Thai rice, or drink coffee from Central America, you are influencing the hydrology of those regions—taking a share of the Indus River, the Mekong River, or the Costa Rican rains. You may be helping rivers run dry.

Economists call the water involved in the growing and manufacture of products traded around the world "virtual water." In this terminology, every ton of wheat arriving at a dockside carries with it in virtual form the thousand tons of water needed to grow it. The global virtual-water trade is estimated to be around 800 million acre-feet a year, or twenty Nile Rivers. Of that, two thirds is in a huge range of crops, from grains to vegetable oil, sugar to cotton; a quarter is in meat and dairy products; and just a tenth is in industrial products. That means that nearly a tenth of all the water used in raising crops goes into the international virtual-water trade. This trade "moves water in volumes and over distances beyond the wildest imaginings of water engineers," says Tony Allan, of the School of Oriental and African Studies in London, who invented the term "virtual water."

The biggest net exporter of virtual water is the United States. It exports around a third of all the water it withdraws from the natural environment. Much of that is in grains, either directly or via meat. The United States is emptying critical underground water reserves, such as those beneath the High Plains, to grow grain for export. It also exports an amazing 80 million acre-feet of virtual water in beef. Other major exporters of virtual water include Canada (grain), Australia (cotton and sugar), Argentina (beef), and Thailand (rice).

Major importers of virtual water include Japan and the European Union. Few of these countries are short of water, so there are ethical questions about how much they should be doing this. But for other importers, virtual water is a vital lifeline. Iran, Egypt, and Algeria could starve without it; likewise water-stressed Jordan, which effectively imports between 80 and 90 percent of its water in the form of food. "The Middle East ran out of water some years ago. It is the first major region to do so in the history of the world," says Allan. He estimates that more water flows into the Middle East each year as a result of imports of virtual water than flows down the Nile.

While many nations relieve their water shortages by importing virtual water, some exacerbate their problems by exporting it. Israel and arid southern Spain both export water in tomatoes, Ethiopia in coffee. Mexico's virtual-water exports are emptying its largest water body. Lake Chapala, which is the main source of water for its second city, Guadalajara.

Many cotton-growing countries provide a vivid example of this perverse water trade. Cotton grows best in hot lands with year-round sun. Deserts, in other words. Old European colonies and protectorates such as Egypt, Sudan, and Pakistan still empty the Nile and the Indus for cotton-growing, as they did when Britain ruled and Lancashire cotton mills had to be supplied. When Russia transformed the deserts of Central Asia into a vast cotton plantation, it sowed the seeds of the destruction of the Aral Sea. Most of the missing water for the shriveling sea has in effect been exported over the past half-century in the form of virtual water that continues to clothe the Soviet Union.

Some analysts say that globally, the virtual-water trade significantly reduces water demand for growing crops. It enables farmers to grow crops where water requirements are less, they say. But this is mainly because the biggest trade in virtual water is the export of wheat and corn from temperate lands like the United States and Canada to hotter lands where the same crops would require more water. But for many other crops, such as cotton and sugar, the trade in virtual water looks like terribly bad business for the exporters.

Pakistan consumes more than 40 million acre-feet of water a year from the Indus River—almost a third of the river's total flow and enough to prevent any water from reaching the Arabian Sea—in order to grow cotton. How much sense does that make? And what logic is there in the United States pumping out the High Plains aquifer to add to a global grain glut? Whatever the virtues of the global trade in virtual water, the practice lies at the heart of some of the most intractable hydrological crises on the planet.

Over the past twenty years, tens of millions of small farmers across much of the poor world have been pumping water from beneath their fields. As in India, it is an extensive revolution born of two factors: first, the failure of government-built irrigation systems that tap rivers to deliver the water the farmers need; and second, advances in technology that allow them to drill far deeper into the earth for water than they could with their old hand-dug wells and to buy cheap Japanese pumps to bring the water to the surface.

There are few official statistics, and there is no way of collecting them. But just three countries—India, China, and Pakistan—probably pump out around 235 million acre-feet of underground water a year from the new tube wells. They account for more than half of the world's total use of underground water for agriculture. And they are living on borrowed time, sucking dry the continent's water reserves.

Every year perhaps 100 million Chinese eat food grown with underground water that the rains are not replacing. There are, as we have seen, another 200 million or so doing the same in India. In the Pakistani province of Punjab, which produces 90 percent of that country's wheat, farmers compensate for diminishing deliveries of water from the Indus River by pumping from beneath their own fields. They pump 30 percent more than is recharged, and water tables are plunging by three to six feet a year.

Overall total pumping in India, China, and Pakistan probably exceeds recharge by 120 to 160 million acre-feet a year. The boom has so far lasted twenty years; the bust could be less than twenty years away. The consequences of the eventual, inevitable failure of underground water in these countries could be catastrophic. But it is a crisis that has not yet registered on the radar screens of governments or aid agencies. When a river runs dry, it is very visible. But underground water is invisible. Only the farmers know they have to drill deeper and deeper to find it. And few in the corridors of power talk to farmers about a slow-burning disaster that will one day affect hundreds of millions of people.

It won't happen everywhere at the same time, of course. Each aquifer has its own countdown to extinction. But even so, as Tushaar Shah puts it, "The overuse of water in Asia's underground aquifers will spell disaster for millions of the region's poor people, who depend on it." And as each aquifer dries up, it will undermine the world's ability to feed itself.

Farmers in other heavily populated countries are beginning to sink tube wells and buy pumps with equal enthusiasm. In the past decade, Vietnamese farmers have quadrupled the number of tube wells to more than a million. Sri Lanka, Indonesia, Iran, and Bangladesh are not far behind. These countries are at the heart of what the agronomist and environmentalist Lester Brown calls "Asia's food bubble." Record farm outputs in recent years, he says, have been made possible only by an unsustainable assault on this fast-diminishing resource. It is a bubble that is bound to burst. "The question is not if, but when," he says.

Asia leads the way in this unsustainable groundwater revolution, but countries such as Mexico, Argentina, Brazil, Saudi Arabia, and Morocco are increasingly significant players. Sub-Saharan Africa is not far behind. As rivers fail, underground sources provide a third of the world's water. By some calculations, as much as a tenth of the world's food is being grown using underground water that is not being replaced by the rains. Major cities—among them Beijing and Tannin, Mexico City and Bangkok—are also growing increasingly reliant on pumping out underground reserves.

As water tables sink, the revolution is putting a premium on finding new underground water reserves—preferably ones that are recharged by the rain. New water is still being found in the bowels of the earth. Chinese scientists recently reported the discovery of unsuspected water beneath a vast system of sand dunes in the Gobi Desert in Inner Mongolia, which they claim is being replenished from mountains to the south. And hydrologists reported new research into the likely size of the Guarani aquifer, which stretches beneath more than 400,000 square miles of Brazil, Argentina, Uruguay, and Paraguay. It may contain 40 billion acre-feet of water—as much as flows down the Amazon River in seven years. And it is still being replenished by the rain. The aquifer is already supplying 15 million people with water without any general decline in water tables. Hydrologists now believe it could one day supply as many as 200 million people. They want to build an aqueduct to take the water to the world's third largest city, São Paulo.

But as fast as one door opens, another closes. The High Plains of the Midwest are part of American history. The first white settlers going west transformed the plains from buffalo hunting grounds into rough pasture for cattle. It was the land of the cowboy. Then came the plow, and pasture became dry prairie until the drought of the 1930s blew the soil away. Since those Dust Bowl days, when millions of sharecroppers abandoned the land and trekked on to California, the arid plains have been transformed once again through the pumping of water from a giant underground reserve discovered beneath them.

Known as the Ogallala aquifer after the Sioux nation that once hunted buffalo here, the reserve stretches beneath most of Nebraska, southern South Dakota, western Kansas, Oklahoma, and Texas and parts of eastern New Mexico, Colorado, and Wyoming. In the 1930s, as the Dust Bowl refugees poured into California, just 600 wells tapped this aquifer. But by the late 1970s there were 200,000 wells, supplying 22 million acre-feet a year to more than a third of the country's irrigated fields.

The aquifer was an enormous U.S. resource, but also a global one. In a good year, the High Plains produced three quarters of the wheat traded on the world market—restocking empty Russian grain stores, feeding starving Ethiopians, and keeping Egyptians fed as the Nile ran dry. The United States became the world's biggest exporter of virtual water—but at the expense of draining the Ogallala.

The problem is that much of the aquifer is to all intents and purposes a fossil resource laid down in wetter times. Little water is added to it from the region's scanty rains. And starting in a few isolated southern pockets, wells have been drying out now for more than thirty years. The first to hear the news, recorded a U.S. water researcher named Sandra Postel, were the farmers of Deaf Smith County, in the Texas panhandle. During the summer of 1970, a well that had been pumping water since 1936 suddenly went dry. And many others have followed.

Today more than a quarter of the aquifer is gone in parts of Texas, Oklahoma, and Kansas, and over wide areas the water table has fallen by more than 100 feet. All well-sinking is banned in some places, and fewer than 10 million acre-feet are now pumped annually, less than half the output in the 1970s. Fly over the land and you can see the circular marks where rotating sprinklers once kept the soil wet and the fields green—but the soil is now dry and the fields are brown. The sagebrush and buffalo grass are returning. The buffalo may follow.

Nationally, the United States is not heavily dependent on underground water. Aquifers provide less than 1 percent of water supplied in New England, for instance. But a third of all its irrigation water does come from underground, and some states in the South and West would be more or less literally lost without it. The country's three major aquifers—the Ogallala, the Central Valley aquifer in California, and the Southwest aquifer—are all in the arid West. The Southwest aquifer in particular is vital. In Arizona, it is virtually the only water source within the state. No wonder that, across the West,

cities are buying up farms to get their rights to pump these aquifers.

But overpumping in the West is as widespread as on the High Plains. For many years Arizona has been removing underground water at twice the rate that rains can replace it. In California people pump out 15 percent more than the rains replenish—an overdraft of 1.3 million acre-feet a year. The combined annual overpumping of the Ogallala, Central Valley, and Southwest aquifers has been 30 million acre-feet, resulting in a cumulative total of over 800 million acre-feet in recent years. And as the aquifers empty, the honeycomb rock that once held the water in their pores are gradually being crushed by the rocks above. One estimate is that this crushing has permanently reduced the water storage capacity of the Central Valley aquifer by half—the equivalent of blowing up a dam the size of Grand Coulee.

The Common Wealth

From the peat bogs of Scotland to the backwaters of the Mississippi delta, from the lagoons of Venice to the flooded forests of Cambodia, and from the frozen tundra of Siberia to the salt lakes of the Australian outback, wetlands are an in-between world. Sometimes wet and sometimes dry, sometimes land and sometimes water, sometimes saline and sometimes fresh, they change their character with the seasons. Their wealth is tied up in hard-to-measure intangibles such as flood pulses and the fertile silts suspended in their waters. They may stay virtually dry for several years before being replenished by violent floods. Their wildlife is similarly transitory, with migrating birds, fish, and even mammals coming and going. And like rainforests, they are generally not owned by anyone. Their rippling waters, shifting land, and migrating wildlife are a common resource available to anyone willing to brave the elements to get them.

These are all unhelpful attributes in the modern world, where certainty in nature is valued more than fluidity, where floods are a "bad thing," where private property is the universal currency, and where resources constantly on the move are hard to own. In every sense, the wealth of wetlands can slip through your fingers. The modern world wants to enclose and privatize and tame them, much as the water in flowing rivers is captured by damming them. The owners of wetlands, despairing of exploiting the moving feast, often prefer to drain it all, fence in the land, and start again. And for generations engineers have regarded any water draining into wetlands as somehow wasted and ripe for diversion to other human uses.

So the world's wet places are being emptied. By some estimates, half of them have gone already. Next to rainforests, they are the most productive ecosystems—for humans as well as for nature. Yet they are disappearing with far less fuss: drained, dyked, canaled, concreted over, turned into shrimp farms and rice paddies, dammed, dredged, and filled with solid waste.

As one recent international research project on wetlands economics observed in the journal *Nature*, "The social benefits of retaining wetlands, arising from sustainable hunting, angling, and trapping, greatly exceeded the agricultural gains" if the wetlands were drained. And yet they are still drained, because the social gains often accrue to people beyond the wetland, while the agricultural gains can be captured entirely by the owner of the land. The study concluded that "our relentless conversion and degradation of remaining natural habitats—including wetlands—is eroding overall human welfare for short-term private gain." As scarce natural resources become rarer, the social benefit from protecting them grows rather than diminishes. Preserving what remains "makes overwhelming economic as well as moral sense."

Let us settle for three examples here, from three of the world's largest wetlands—the Sudd swamp on the Nile River in Sudan, the Pantanal in the heart of South America, and the Okavango delta in southern Africa. Two of these involve plans for canals that would rush waters past the wetland, in one case to reduce evaporation, in the other to aid navigation. The third is an apparently modest attempt to tinker with a virgin ecosystem for laudable development goals, but it could have huge unintended consequences. In all three cases, the potential beneficiaries are remote from the wetland and have little to lose by its demise. In each case, too, a common resource of global importance is at risk of being destroyed for narrow self-interest.

The Sudd is the world's second largest swamp. It occupies a swath of southern Sudan on the White Nile, one of two great tributaries of the Nile River. Places do not come much more remote than this, and the Sudd is a formidable natural obstacle. It prevented navigation upstream by generations of European explorers seeking one of the Holy Grails of geography, the source of the Nile.

The Sudd is a natural wonder of the world. Its numberless channels contain an ever-shifting maze of papyrus islands, the *sudds* (Arabic for blockages) that give it its name. *Sudds* can be up to half a mile long and many yards thick—thick enough to carry large herds of hippos and elephants. Sometimes they block the whole channel until, like a dam, they burst. This must be an extraordinary sight. One British colonial officer on a rare nineteenth-century foray into the swamp described how, after such a burst, "crocodiles were whirled round and round, and the river was covered with dead and dying hippos." Others were not so enthused by the Sudd. The Victorian water engineer Sir William Garstin said of it, "No one who has not seen it can have any real idea of its supreme dreariness and its utter desolation. To my mind the most barren desert is a bright and cheerful locality compared with the White Nile marshland."

The Sudd is a 300-mile turnout on the long journey of the White Nile's waters from Central Africa to the Mediterranean. The river's water takes more than a year to pass through the swamps, during which time half its volume, some 4 million acre-feet, is lost to evaporation. Those are acre-feet that people downstream have long wanted to get their hands on. It was the unenthusiastic Garstin who first proposed either dredging the Sudd or diverting the White Nile away from its suffocating embrace in order to preserve its water from the sun. He never got his way, but modern engineers have not lost the urge. Egypt, which is already using every drop of the Nile that flows through its land, wants to turn dreams into action. And that is why for the past thirty years a vast Rube Goldberg machine called a "bucketwheel" has been sitting in the middle of the bush of southern Sudan.

The bucketwheel is the canal diggers' answer to the combine harvester. It is a giant laser-guided digging wheel weighing 2500 tons and as tall as a five-story building, a fearsome parody of an ancient Persian waterwheel with twelve buckets attached to the wheel to grab earth rather than water. It was in fact constructed in the 1960s to dig a canal across the Punjab in Pakistan. After that job was done, it was dismantled and carried by truck, train, ship, and finally camel to the edge of the Sudd. Its new task was to dig a channel 160 miles long and 160 feet wide to allow the Nile to skirt the eastern side of the marshes.

In June 1978, the bucketwheel began excavating what was called the Jonglei Canal. The giant wheel rotated once every minute, night and day. At each revolution its buckets ate up and threw aside enough Nile sand to fill an Olympic swimming pool. The machine consumed 10,500 gallons of fuel a day—more than all the buildings in Juba, the regional capital.

This, however, was not a good moment to dig a canal through the backwoods of southern Sudan.

The route passed through the land of Dinka cattle herders. They were seriously inconvenienced by the canal, which cut off their lands from dry-season pastures on the fringes of the Sudd. Moreover, the Dinka and other groups in the non-Muslim south of the country were about to embark on a war of liberation from their Muslim masters far to the north in Khartoum.

By February 1984 the bucketwheel had carved out a third of the route, at a cost of $100 million. Then the Sudanese People's Liberation Army struck the contractors' camp. They destroyed everything except the great machine itself and ran off with foreign hostages. This was no ignorant piece of terrorism. Its perpetrator, the leader of the SPLA, was John Garang, a young Dinka tribesman who had received an American university education in the 1970s. While at Iowa State he had written a doctoral dissertation on the iniquities of the Jonglei plan. In it, he argued that the canal would suck southern Sudan dry of its greatest resource—the waters of the Sudd.

Though the hostages were released unharmed a year after their capture, the contractors never returned to their project. Civil war raged around the Sudd for twenty years, and the bucketwheel has sat, more or less intact, beside the wetland ever since. At the end of 2004, Garang and the authorities in Khartoum concluded peace talks that provided a fair degree of autonomy for southern Sudan. Insiders said that twenty years on, Garang was better disposed toward the Jonglei project and might finally consent to the construction of the canal, with proper safeguards for his people. After his death in an air accident in 2005, it was far from clear what the consequences might be for the Jonglei. But Egypt was certainly eager to put up the cash. The new millennium saw it engaged in a giant new Saudi-funded irrigation project in the western desert, known as the Toshka Lakes scheme, which will eventually require 4 million acre-feet of water—coincidentally, precisely the anticipated yield of the Jonglei Canal.

The canal may now be dug. However, in their concern to end evaporation of Nile water from the Sudd, the Egyptians seem to have lost sight of an even bigger evaporation loss far closer to home. Though it is rarely mentioned, roughly three times more water evaporates every year from the surface of Lake Nasser, the vast reservoir behind Egypt's Aswan High Dam, than from the Sudd. This Russian-built symbol of Egyptian independence, the crowning glory of the rule of Colonel Abdel Nasser in the 1960s, loses roughly 12 million acre-feet from its huge surface every year. In a dry year, that is more than a third of the river's entire flow. But nobody is suggesting that the reservoir should be emptied.

Mekong: Feel the Pulse

The diminutive king and queen of Cambodia appeared on the balcony of the royal palace overlooking the Tonle Sap River. Like fairy-tale monarchs, Norodom Sihanouk and his queen, Monineath, serenaded their subjects—upwards of a million of them—gathered on the riverside promenade of the beautiful old colonial capital of Phnom Penh. It was the high spot of the Cambodian year, and the audience between monarch and subjects was part of one of the world's oldest boat-racing festivals, the Bon Om Touk.

That year, 2003, around four hundred boats, each decorated with paintings of water serpents, rushed two by two down the river to a finish line right by the royal palace, where the Tonle Sap flows into the mighty Mekong. Some boats contained seventy frenetic oarsmen, all standing in a line, pounding the water from the narrow vessel. Altogether, thousands took part in the races, while a million more Cambodians, from across one of the poorest and least

urbanized nations on earth, flooded into the city for a weekend of eating and camping on the riverside.

This festival has taken place since the twelfth century, always at the full moon in late October or early November. It is a celebration of one extraordinary fact about the river on which it takes place. The Tonle Sap is one of the few rivers in the world that reverses its flow. It does it every year, right in front of the palace.

It happens because every year, at the start of the monsoon rains, the often quiet Mekong becomes a raging torrent that overwhelms its tributary, forcing it to surge back upstream for some 125 miles. At the height of the monsoon, this reverse flow swallows a fifth of the Mekong's prodigious waters, and the tiny Tonle Sap is for a few months one of the world's great rivers—albeit flowing backward. The river backs up into a great lake that itself spills over its banks, flooding a wide area of rainforest, which turns into a vast fish nursery for the Mekong in the heart of Cambodia.

For five months fish grow in profusion in the silty forest waters, before being washed back into the main river as the monsoon abates and the Tonle Sap resumes its proper flow. The annual boating festival marks this unique spectacle and celebrates the great flush of fish that leaves the forest with the water. It is a flush that, in the months to come, will feed tens of millions of people up and down the Mekong River system.

Or that is how it should be. But in 2003 there was consternation on the river. The Mekong floods had been the poorest on record, the reversal of the Tonle Sap's flow had been short and tentative, and the fishing had been terrible. Some in the crowd said that mysterious changes far upstream on the Mekong had neutralized the flood. Fears grew that the reversal of the Tonle Sap might be about to come to an end. Something else was afoot too. In the royal palace. This turned out to be the eighty-year-old king's last serenade of his people. In failing health, he abdicated a few months later. Was this, his subjects asked, the end of an era? The end even of the great flood?

Once the world's rivers teemed with fish. Then, during the twentieth century, most of the rivers were barricaded by dams and their wild flows were tamed. Almost everywhere this has caused a drastic decline in wild fisheries. Natural fish nurseries have been wrecked and fish migrations disrupted by the unnatural flow rhythms created by the dams. But on the Mekong, the great artery of Southeast Asia, half a century of warfare has kept the dam builders away, and the natural seasons of the river have continued uninterrupted. The river has remained free to trespass over its banks and flood forests in summer, then retreat to small pools and rapids in winter. The fish have prospered. And not just the fish, for the fishing communities have enjoyed a bonanza now extremely rare in the world. In Cambodia, where the river flood is most intense, the people live off wild freshwater fish to an extent unknown anywhere else. Even the poorest can dine like kings.

The Mekong remains true to its popular image as the "sweet serpent" of Southeast Asia. It winds for 2800 miles out of the ice fields of eastern Tibet and through a long series of deep gorges in the mountains of southern China before tumbling down rapids to flood the rainforests of Laos and Cambodia and sliding into the sea through its delta in Vietnam. The Mekong is far from being the world's largest river. Its average annual discharge of 380 million acre-feet makes it fourteenth in the riverine pecking order. But its flow is today the most variable of those of the major world rivers. During the summer monsoon, it contains up to fifty times more water than in the long dry season. Then it has the third biggest flow of any river in the world, exceeded only by the Amazon and the Brahmaputra.

"The Mekong is not just another river," says Chris Barlow, a fisheries researcher at the Mekong River Commission, an intergovernmental science agency. "It is the least modified of all the major rivers in the world. Animals have evolved to exploit its flood pulse, and local societies have developed that way, too." Four fifths of the population of Cambodia is involved in fishing and processing the harvest. Some 60 million people in the Mekong's lower basin—in Laos, Cambodia, and Vietnam—draw their food and income from the river and its wetlands, catching some 2 million tons of fish a year. That is over 2 percent of the entire world catch of wild fish from both rivers and the sea. Among rivers, only the Amazon produces more.

The Mekong is a salutary lesson in what the world has lost since it began erecting concrete barriers across nature's finest. More than a thousand species of fish live in the Mekong, again more than anywhere except the Amazon. Hundreds of species are regularly caught and eaten, but the most important is the humble trey riel, a small sardinelike relative of the carp that turns up in almost every fishing net on the river. It is, says Joern Kristensen, the head of the river commission, "both meat and milk," because it provides both protein and calcium. Oxfam says that the river fisheries of Cambodia "make a bigger contribution to economic well-being and food security than in any other country." Cambodians are more dependent on wild protein than the people of almost any nation on earth. As a result, one of the world's poorest countries is one of its best fed.

* * *

Forty years [after the Vietnam War], Southeast Asia is finally at peace, and demand for water and electricity is soaring. For engineers, the river is the obvious source for both. Thailand, Vietnam, and Laos have already built dams on tributaries of the Mekong, and hydroelectricity is now the biggest legal export from Laos. Most of these dams have proved problematic in one way or another. Vietnam's blocking of the Se San at Yali Falls is causing chaos as unannounced releases of water flood villages over the border in Cambodia. Thailand's Pak Mun Dam on the Mun River wrecked local fisheries. The Nam Theun II, for which Laos wants international funding, has become a cause célèbre before a stone has been laid, because locals fear it too will kill their fish.

China, meanwhile, is building a cascade of eight huge dams on the main stem of the Mekong, which it calls the Lancang. The first two are already operating and generating concern downstream all the way to the sea. As the turbines are switched on and off to meet changes in demand for power and their reservoirs empty and fill, water levels in the river fluctuate by up to 3 feet a day for hundreds of miles downstream. Peter-John Meynell, of the World Conservation Union, who regularly sails on the river, says, "I feel the wash myself going down the river. Even big boats find it difficult because of the surges from the dams. Local fishers are losing their livelihoods as a result."

Meanwhile, the initial filling of the first two dams—the Manwan in 1993 and the Dachaoshan in 2003—coincided with unusually low flows on the Mekong all the way down to the Tonle Sap, and with poor fish catches. In other years, hydrologists have measured a reduction in the Mekong monsoon flow that they suspect is due in part to the dams holding back the floodwaters for release through the dry season. The Mekong's pulse is being weakened.

But this is just the start. Construction on a far bigger dam began in 2002. The Xiaowan Dam will tower 958 feet over the river, as high as the Eiffel Tower. Its reservoir, at 105 miles long, will be twenty times the size of the two existing reservoirs combined and second in China only to the Three Gorges, on the Yangtze. After that will come an 800-foot-high dam, the Nuozhadu, which will have an even bigger reservoir. By early next decade, the cascade of dams will be able to store 32 million acre-feet of water—more than half the river's flow as it leaves China.

According to Chinese operational plans, the river's dry-season flow out of China will triple after the cascade is finished, and the huge wet-season flow will fall by a quarter. The effect will diminish as you go downstream, but with 40 percent of the river's flow coming from China in the dry season, it will still be enough to halve the strength of the flood pulse in Cambodia as it passes the royal palace. The effect of that on the reversal of the Tonle Sap is harder to predict. In a worst-case scenario, though, the reversal could give out altogether.

Changing Climate

[Global warming] will almost certainly increase rainfall enough to fill twenty Niles. This will almost certainly increase global rainfall. Also, shifting trajectories of rain-giving climate systems, like Atlantic cyclones, will mean that the total rainfall will be redistributed. Many middle latitudes will become drier. Meanwhile, the higher temperatures will also mean faster evaporation of water on land, so soils will dry out more quickly. That means less of the rainfall will reach rivers.

Future flows on individual rivers will depend in large part on the balance between the competing effects of changing precipitation patterns and great evaporation. The rule of thumb seems to be that dry areas will become drier while wet areas will become wetter. Globally, climate will become more extreme, and rivers will respond in kind.

As a result, many of the rivers that provide water in the world's most densely populated areas and where water is already in the shortest supply will be in still deeper trouble soon. In northeastern China, the savanna grassland of Africa, the Mediterranean, and the southern and western coasts of Australia, rains will probably diminish, evaporation will certainly be greater—and the rivers will run dry.

The U.S. government's Scripps Institution of Oceanography estimates that reservoir levels in the Colorado will fall by a third as declining rainfall and rising evaporation combine to reduce moisture by up to 40 percent across the southern and western states. The Niger River, which waters five poor and arid countries in West Africa, is expected to lose a third of its water, and the Nile the lifeblood of Egypt and Sudan, could lose a fifth. Inland seas will be at special risk as many rivers in continental interiors lose flow. Lake Chad has almost succumbed and may suffer further in future. Also under threat are the Caspian Sea and Lake Balkhash in Central Asia, Lakes Tanganyika and Malawi in East Africa, and Europe's largest, Lake Balaton, in Hungary.

In contrast, giant tropical rivers like the Amazon and the Orinoco in South America and the Congo in Africa will become even more bloated than today, the models predict. Similarly, the great Arctic rivers of northern Canada and Siberia will probably gain water as warmer air holds more moisture and more rain falls on their catchments. So the Mackenzie and the Yukon in Canada and the Ob, the Yenisei, and the Lena in Siberia will rage even more fiercely—40 percent more, according to one big study.

There is some evidence that these trends are already well under way. In early 2005, the U.S. government climatologist Kevin Trenberth showed that after a century of little change, there has been a surge in instances of severe drought around the world since 1970. The proportion of the earth's land surface suffering very dry conditions rose from 15 percent then to 30 percent at the start of the twenty-first century, he said. Meanwhile, scientists from the Met Office reported that "far northern rivers are discharging increasing amounts of freshwater into the Arctic Ocean due to intensified precipitation caused by global warming." There has been an annual increase in river flow into the Arctic of 7.3 million acre-feet since the 1960s, they found.

A British study in the Himalayas found that the Indus could increase its flow by between 15 and 90 percent in the first half of this century as glaciers melt, and then decrease to between 30 and 90 percent of the current flow later in the century. Likewise, American researchers believe that the spring melt-waters from the Sierra Nevada snowpack, which today sustain summer irrigation of crops and lawns across the desert lowlands of California, could diminish by 70 to 80 percent over the next fifty years. As the rushing meltwaters become a trickle, some of the most productive agricultural lands in the world could dry up.

One important effect of existing glaciers is to stabilize river flow between years. They absorb highly variable monsoon rains and provide a strong, regular flood pulse in the summer melting season. This is vital to people living downstream. The glaciers of the Himalayas and Tibet feed seven of the greatest rivers in Asia—the Ganges, Indus, Brahmaputra, Salween, Irrawaddy, Mekong, and Yangtze—ensuring reliable water supplies for 2 billion people. But in half a century or so, the glacier flows in many of these rivers will dwindle and be replaced by much more fickle flows from rain in the mountains. That is a serious threat to Asia's future.

"Once the glaciers go, you're down to whatever happens to fall out of the sky," says the British glaciologist Martin Price. And not just in Asia. Down the Andes, cities like La Paz, Lima, and Quito—the capitals of Bolivia, Peru, and Ecuador, respectively—depend on glaciers for secure water and hydroelectric power, but the glaciers are disappearing fast.

Uncertainty, then, is the greatest difficulty. Some of these predictions assuredly will not come true. Equally certainly, there will be real-life horror stories unconsidered in the climate models. This creates huge dilemmas for water engineers. How can they design dams and irrigation schemes, with expected lifetimes of fifty or a hundred years, to cope with such uncertainty? In all probability, they can't.

Wonders of the World

Modern water engineering began in earnest with the Hoover Dam—the first superdam. The 700-foot plug across the Boulder Canyon on the Colorado River was completed in 1935. It was taller than a sixty-story building, bigger than the Great Pyramid of Egypt, and contained enough concrete to pave a highway from San Francisco to New York. Behind it grew Lake Mead, which could hold more than twice the river's annual flow. As Francis Crowe, the BuRec surveyor on the project, later put it, "I was wild to build this dam—the biggest dam built by anyone, anywhere." This was not the language of technical reports, but it was perhaps the truest reflection of the motivation behind much large dam construction on the Colorado and around the world.

The white concrete of the Hoover Dam became a symbol of Franklin Roosevelt's New Deal public works projects, and of a broader lust to remake the landscape. It was soon joined by the equally talismanic Grand Coulee Dam on the Columbia River. Famed folksinger Woody Guthrie even wrote a collection of songs, the Columbia River Songs, for the Columbia River and the Grand Coulee Dam; in it he called them the newest wonders of the world.

These dams ushered in a postwar world in which dams became symbols of modernism, of economic development, and of mankind's control over nature. Russia wanted its own Hoover Dam; so did Egypt and Japan and China and India. Hoovers sprouted across Latin America; Britain built replicas for its colonies as parting gifts before independence. No nation-state, it seemed, was complete without one.

"Except for global warming, there has been no more drastic human alteration of the landscape in the last fifty years than the damming, regulation, and diversion of the world's rivers," Beard told the Japanese demonstrators. But he was not boasting; he was apologizing. "Those who promoted dam projects were not honest about costs and benefits. The truth is that dam proponents would say just about anything to get a dam project approved."

The decisions to build were "political, benefiting particular politicians or their benefactors rather than solving a problem. In our experience at BuRec, the actual total costs of completing projects exceeded the original estimate typically by 50 percent. And the actual contribution made to the national economy by these dam projects was small in comparison to the alternative uses that could have been made with the public funds they swallowed up. We are now spending billions of dollars to correct the unanticipated impacts such as lost fisheries, salinized soils, and desiccated wetlands." Ouch.

Rivers today are worth more in America as amenities for fishing and tourism than as water for filling reservoirs, Beard said: "We've started tearing down dams." Is the United States a special case? Far from it. "In my view, we are starting down a similar path throughout the world," he added. "The time when large dam projects are a realistic answer to solving water problems is behind us."

Like Beard, modern environmentalists have come to see large dams as engines of environmental destruction. But it was not always so. For many years, environmentalists in both North America and Europe supported the dam builders. Rather like wind power today, hydroelectricity was seen as a new, clean, cheap source of electricity. In Europe, concern that dams might interfere with the natural flow of rivers such as the Danube, the Rhine, and the Rhone was tempered by the fear that the alternative was pollution from fossil-fuel power stations. As late as 1996, the head of WWF (the World Wildlife Fund) in Austria told me, "Even though we love the mountains, most environmentalists in Austria still support the construction of dams in their valleys."

Hydroelectric dams do generate a huge amount of electricity—around a fifth of the global total. More than sixty countries depend on them for more than half their power. One dam, at Itaipu on the Parana River between Brazil and Paraguay, has a generating capacity of 12,600 megawatts—equivalent to a dozen conventional power stations. It supplies São Paulo and Rio de Janeiro two of the world's megacities. China's Three Gorges on the Yangtze will be even bigger. Most modern dams serve double or triple functions, also supplying water for irrigation projects and city faucets and sometimes claiming flood prevention capabilities as well.

The world's large dams now hold more than 5.5 billion acre-feet of water. Most big river systems, including the twenty largest and the eight with the most biological diversity—the Amazon, Orinoco, Ganges, Brahmaputraj, Zambezi, Amur, Yenisei, and Indus—all now have dams on them. Most of the surviving untamed rivers of the world are in the empty Arctic tundra and northern boreal forests. The largest is the Yukon, in northern Canada. The handful in tropical lands are rapidly becoming extinct. The Salween, which runs for some 1600 miles from China through Burma and Thailand, is about to be dammed for the first time. On rivers like the Colorado, the Volta in West Africa, and the Nile, the big dams can hold two or three times the actual annual flow. And yet they remain an essentially experimental technology. Their hydrological, ecological, and social effects have been huge. But for many years their status as symbols of modernism insulated them from serious appraisal. Even in the 1990s, fewer than half of all proposed dams had an environmental impact appraisal before construction began. And even fewer had the consent of the people they displaced. Only since the late 1990s have serious steps been taken internationally to establish whether their benefits outweigh the environmental, social, and economic costs.

The focal point for this reevaluation has been the World Bank. In the second half of the twentieth century, the bank spent an estimated $75 billion on building large dams in ninety-two countries. But by the late 1980s, its own cost-benefit analyses

questioned the value for money represented by this investment bonanza. They catalogued huge cost overruns, billion-dollar corruption scandals, poor design and bogus hydrology that left reservoirs empty, turbines that were never connected to national grids, and promised downstream irrigation projects that never got built at all. Bank-financed dams, the analysts also discovered, had caused the forced resettlement of 10 million people.

Amid a rising chorus of opposition around the world, the bank pulled out of contributing to the funding of a high-profile dam on the Narmada River in India. In a quandary over how to proceed, it appointed a World Commission on Dams to assess the successes and failures of large dams and come up with some ground rules for what a successful dam project might look like.

The final report from the commission was launched in a blaze of publicity in London in late 2000 under the benign gaze of Nelson Mandela and the rather sterner visage of the World Bank's president, James Wolfensohn. It was even more scathing about large dams than the bank could have feared at the outset. It endorsed many of the environmentalists' most trenchant criticisms. Most dams just don't deliver as advertised, the commission said. Average cost overruns were 56 percent. Half of hydroelectric dams produced significantly less power than promised; two thirds of those built to supply water to cities delivered less water than promised; a quarter of them delivered less than half what their brochures claimed. Dams built to irrigate fields were no better. A quarter of them irrigated less than 35 percent of the land intended. Even dams that promised to protect against floods "have increased the vulnerability of river communities to floods," often because their reservoirs have been kept full to maximize hydroelectric production.

And dams have taken huge amounts of land—land on which people once lived. All told, at least 80 million rural people worldwide had lost their homes, land, and livelihoods, the commission found. The Akosombo on the Volta in Ghana expelled 80,000 people, the Aswan High in Egypt 120,000, the Damadur in India 90,000, the Kariba in southern Africa almost 60,000, the Tarbela in Pakistan more than 90,000. And for what? My own estimates show that the Aswan High Dam generates just 2 kilowatts of electricity for every acre of land flooded; Kariba generates just 1.2 kilowatts, Akosombo 0.36 kilowatts. Others are even worse: the Kompeinga Dam in Burkina Faso generates just 0.28 kilowatts per acre, and Brokopondo in Suriname is worst of all, at 0.08 kilowatts.

Dams' destruction of ecology has been extensive. The commission found that far from "greening the desert" as promised, many dams may have encouraged its advance by desiccating wetlands and delivering salt to fields. A quarter of the world's irrigated land, much of it watered by dams, has been damaged by salt and waterlogging. Meanwhile, accumulations of silt have been reduced by more than half the storage capacity behind a tenth of older dams. And by stopping the flow of silt downstream, dams have universally reduced the fertility of floodplains and "invariably" caused erosion of riverbanks, coastal deltas, and even distant coastlines. Coastal lagoons are being washed away all along the West African coast owing to dams.

A study published early in 2005 found that the world's wild rivers are rapidly becoming extinct. Of the three hundred largest river systems, almost two hundred, including all of the twenty largest, now have dams on them—and engineers are rapidly moving onto the last untamed flows to tap their waters for hydroelectricity or irrigation.

Most of the remaining untamed rivers are in the empty Arctic tundra and northern boreal forests. The largest surviving wild river system is the Yukon in remote northern Canada, the world's twenty-second largest river by volume. Europe's last three undammed river systems are all in northern Russia. But Christer Nilsson, a landscape ecologist at the University of Umea in Sweden, who led the research, said that many dry parts of the world have become "totally devoid of unaffected river systems." And surviving rainforest rivers are under increasing attack. Rivers about to be dammed for the first time include the Salween, which runs for 1500 miles from China through the jungles of Burma and Thailand; the Rajang, in Malaysian Borneo; the Jequitinhonha in Brazil; the Ca, in Vietnam; and the Agusan, on the Philippine island of Mindanao.

By interrupting natural river flows, dams have wrecked fisheries from the Columbia to the Ganges. Fish stocks established in the reservoirs behind dams rarely come close to compensating. The Grand Coulee and its fellow dams on the Columbia River have destroyed one of the world's largest and most lucrative salmon fisheries. The fish, according to one study submitted to the commission, would have been worth more than the electricity generated by the dams.

Usually the benefits are short-term while the costs are long-term. Usually urban elites gain most, especially from hydroelectricity, while the poor in the countryside lose most, as their fields are flooded or the rivers and wetlands on which they depend are wrecked. Water, as they say in the American West, flows uphill to money. That is why farmers and fishermen everywhere in the world are being forced to give up their water resources to cities and industry.

It is perhaps no surprise that, notwithstanding the idealism of the early days, autocratic, corrupt, and militaristic governments have come to like dams best. Marshall Goldman, the analyst of Soviet Russia, identified "an almost Freudian fixation … Nothing seems to satisfy the Soviets as much as building a dam." Communist China, with 22,000 large dams, has almost 50 percent of the world total. Spain, under the Fascist leadership of General Franco, built more dams than any nation of comparable size on earth. If nothing else, dams have proved an exceptionally effective technology for turning the unruly flow of rivers into private or state property.

Sun, Silt, and Stagnant Ponds

You could fill every faucet in England for a year with the amount of water that evaporates annually from the surface of Egypt's Lake Nasser. This is one of the more staggering statistics I uncovered while researching this book. According to Egyptian hydrologists, between 8 and 13 million acre-feet of water disappear from the surface of the great reservoir behind the Aswan High Dam—that is, a quarter of the average flow of the river into the reservoir, approaching 40 percent in a dry year.

This huge waste of water in a country dependent for its survival on the Nile should come as no great surprise in Cairo. In the early twentieth century, British colonial engineers forcefully advised against building a dam in the Nubian desert for precisely this reason. It would be far better, they said, to capture water farther upstream, in Ethiopia, where most of the river's water rises. There, a cooler, cloudier climate would reduce evaporation rates and the steep valleys would reduce the surface area of the reservoir. But in the 1950s, when Egypt gained independence and could do what it liked, its leader, Gamal Abdel Nasser, decided on a dam on Egyptian territory, whatever the water losses. He created the second largest manmade lake in the world, with water spreading for 300 miles along the flat Nubian desert deep into Sudan.

Lake Nasser is not alone in such losses, of course. Right across the tropics and beyond, evaporation is a major drain on reservoirs. In the American West as in Egypt, more than 6 feet of water evaporate annually from reservoirs like Elephant Butte, on the Rio Grande, and Lakes Mead and Powell, on the Colorado. A tenth of the flow of the Colorado River evaporates from Lake Powell alone. A typical reservoir in India loses 5 feet. In the parched Australian outback, the losses can exceed 10 feet a year.

What proportion of the reservoir's water supplies this represents depends on the ratio between surface area and capacity and the amount of time the water spends in the reservoir. Like the Aswan High, the Kariba Dam on the Zambezi loses around a quarter of its annual inflow, which is more water than is consumed annually in Zimbabwe. In Namibia analysts estimate that the amount of water evaporating from the Epupa reservoir annually could supply the capital, Windhoek, for forty-two years.

Among the worst offenders must be the Akosombo Dam in Ghana, which holds back a larger surface area of water than any other—approaching 4000 square miles. An evaporation rate of 5 feet would lose 12 million acre-feet a year, which is roughly half the reservoir's input from the Volta River in an average year. Rainfall into the lake may compensate for some of that, but the rain would fall anyway, whereas the evaporation would not happen without human intervention.

Could this evaporation be prevented? Technologists have come up with an ultra-thin layer of organic molecules that can reduce evaporation on small reservoirs by up to a third. But on larger reservoirs, wind breaks the thin surface. And the ecological effects of cutting off the exchange of gases such as oxygen and carbon dioxide between air and water remain largely unknown. So for now we are stuck with the problem.

Igor Shiklomanov, a Russian hydrologist, estimates that 1 quart in every 20 drawn from rivers for human use disappears in reservoir evaporation. That works out at 285 million acre-feet a year, of which 40 percent is lost in Asia, a quarter in Africa, and a sixth in North America. In Australia, 20 million people lose something like 3.2 million acre-feet of water a year—or 53,000 gallons a head. Peter Gleick, the author of a biennial report on the world's water, estimates that an average U.S. hydro-electric dam loses a third of an acre-foot of water per year for every person supplied with electricity. Of course, reservoir evaporation may cause rainfall somewhere else. But even so, it is a lot of water to mislay.

Fetid, choked with weeds, and swarming with mosquitoes, the Balbina reservoir in the Amazon rainforest 100 miles north of Manaus is a billion-dollar boondoggle. The dam rises 150 feet above the forest floor on the Uatuma River, a tributary of the Amazon. The reservoir floods an area forty times the size of Manhattan, but much of it is less than 15 feet deep. Philip Fearnside, from Brazil's National Institute for Research in Manaus, has counted 1500 islands and "so many bays and inlets it looks rather like a cross-section of a human lung."

Even the introduction of a herd of grazing mana-tees has failed to stanch the spread of weeds across the surface. Stagnant water slips through the reservoir's flooded forest for years before reaching the dam's hydroelectric turbines, which have a paltry generating capacity of 112 megawatts. The reservoir needs to flood the equivalent of a soccer field to de-liver enough power to run a small air-conditioning unit back in Manaus.

Even in the Amazon rainforest, that sounds like a waste of land. But the true insanity of this hydroelectric dam has only recently emerged. The rotting vegetation in the flooded forest is producing huge amounts of methane, one of the greenhouse gases thought to be responsible for global warming. The reservoir was created in the 1980s to provide pollution-free electricity for the capital of the Amazon, but by Fearnside's calculations, it produces methane with eight times the greenhouse effect of a coal-fired power station with a similar generating capacity. This is not green electricity.

Balbina is not alone. Brazil is largely powered with hydroelectricity, and Marco Aurelio, of Cidade University in Rio de Janeiro, says that up to half of

Brazil's hydroelectric reservoirs warm the planet more than an equivalent fossil-fuel power plant. The World Commission on Dams warned that greenhouse gases bubble up from every one of the reservoirs in the world where measurements have been made. While probably only a handful are emit-ting more than fossil-fuel power stations, almost all make a significant contribution to atmospheric concentrations. "There is no justification for claim-ing that hydroelectricity does not contribute signifi-cantly to global warming," the commission said.

Vincent St. Louis, of the University of Alberta in Canada, has tried to calculate the global effect of all this methane. He says that reservoirs produce a fifth of all the manmade methane in the atmosphere and make up 7 percent of the manmade greenhouse effect. That is a bigger impact than, for instance, aircraft emissions. His calculation is controversial but persuasive. Until recently, scientists investigat-ing the phenomenon believed that the gases came mostly from vegetation trapped underwater when the reservoir filled. They reasoned that the rotting vegetation would soon be gone and emissions would cease. Not so, it turns out. As reservoirs age, most continue to produce substantial quantities of methane.

Why? For one thing, rotting can be very slow. It takes up to five hundred years for a tree to rot in a stagnant Amazon reservoir. For another, a lot of the rotting vegetation does not come from the reservoir but floats down the rivers that drain into the reser-voir. As long as the reservoir continues to flood, the vegetation will continue to arrive and the reservoir will continue to give off greenhouse gases. Of course, most of this vegetation would have rotted anyway. But without reservoirs, says St. Louis, the decompo-sition would most likely occur in a well-oxygenated river, producing carbon dioxide—whereas tropical reservoirs usually contain little oxygen, and as a result they generate methane instead. Methane is twenty times more potent as a greenhouse gas than carbon dioxide. Reservoirs thus change the way significant amounts of the earth's vegetation rots, and with it dramatically raise the greenhouse effect of the rotting.

This could be political dynamite when the Kyoto Protocol on climate change is discussed. The inclusion of reservoir gases would transform the estimated emissions of greenhouse gases for some nations. If St. Louis's average figure for emissions from tropical reservoirs holds for the Akosombo Dam, the reservoir must be emitting five times as much greenhouse gas annually as all of Ghana's fossil-fuel burning.

The tiny South American nation of French Guiana is in a similar position. Once thought of as one of the world's most greenhouse-friendly nations, French Guiana has a small population, and its industrial emissions are minuscule. But a new dam built in the jungle to power the launch site for Europe's Ariane space rocket is a greenhouse boondoggle on the scale of Balbina. It produces three times as much greenhouse gas as an equivalent coal-burning power station. As a result, French Guiana's real per capita emissions of greenhouse gases are three times those of France and greater even than those of the United States.

Reservoirs are not permanent structures—or not permanently useful, anyway. However well they are managed, they accumulate silt brought down from the headwaters of the rivers they trap. The extreme case is the Yellow River, the world's siltiest river, which succeeded in filling the Sanmenxia reservoir in just two years. China's dam engineers are more sophisticated about how they manage silt flows these days, but the reservoirs on the river currently still have a half-life of less than two decades.

No other river beats this, but most rivers flowing out of the Himalaya heartland of Asia carry substantial silt flows. Many dams on these rivers are likely to be as good as useless in forty or fifty years. The reservoir behind the Tarbela Dam, the largest and most upstream on the Indus River, is now more than a quarter full of silt. An island of silt has already broken the surface and is moving ever closer to the dam itself. By 2025 the reservoir will be three quarters filled and effectively useless. Pakistan is looking for a site for a replacement.

Overall, the world's reservoirs are thought to be losing their storage capacity at a rate of at least 1 percent a year. In China, which has more reservoirs than any other country, the figure is 2 percent. That is a loss of millions of acre-feet a year. "This loss should be of the highest concern for governments across the globe," says Rodney White, of HR Wallingford, a British consulting firm. "The world requires between three hundred and four hundred new dams every year just to maintain current total storage."

In theory, clogged reservoirs could in future be replaced with new ones. The trouble is that most of the world's best potential sites for dams are already taken. Any replacement for the Tarbela will be only a fraction as effective. Future dams will get second and third and fourth best, with less water storage and less hydroelectric potential. And their ecological impacts are likely to be much greater. However good the current crop of dams have been for the world, their successors will flood much more land for far less benefit—and be very substantially more expensive.

Swords of Damocles

Southern Doda in Indian Kashmir is bandit country. Pakistan claims the region, and Muslim youths regularly cross the border into Pakistan for training, returning as guerrillas to fight Indian security forces. The war has been going on for more than a decade, with an estimated 50,000 dead. It usually hits the headlines only when foolhardy Western tourists get caught up in it. But some people believe that the world's first nuclear exchange could one day be triggered between India and Pakistan because of events up here in the foothills of the Himalayas. If so, it may not be over guerrillas or terrorists or even border incursions by regular soldiers. It could be over water, for Kashmir is the gateway through which Pakistan receives most of its water along tributaries of the Indus River.

Pakistan's 150 million people would be in trouble without the Indus. It runs the length of their country. Its waters irrigate most of their crops and generate half their electricity. But the Kashmir gateway is an Achilles heel. India and Pakistan have been in armed conflict three times since the two states were formed on British withdrawal from the subcontinent in 1947, and the first conflict arose when India intervened in Kashmir to cut the flow of tributaries of the Indus on which Pakistan relied. Some in Pakistan fear that India may be about to do the same again.

After a decade of negotiations and skirmishes following the first conflict, the World Bank brokered the 1960 Indus Waters Treaty between the two countries. The treaty bound them to share the flow of the river, with each taking water from three tributaries. The flow of the Chenab River was among those going through Kashmir that were given to Pakistan, and it has since been the biggest source of water for Punjab, the breadbasket of Pakistan. But geography dictates that India always has the potential to stop Pakistan's water. And now, in the heart of southern Doda, Indian engineers are building a dam on the Chenab.

Pakistan sees the 525-foot-high, billion-dollar Baglihar barrage, rising from the bed of the Chenab just before it crosses the border, as a clear breach of the treaty. India denies any breach. The dam is intended only to generate hydroelectricity, it says. It will not remove any water from the river, only channel its discharge downstream through turbines. India says Pakistan will get all the water it should. But Pakistan says that India could, in a future crisis between the two nations, use the barrage to hold up the river's flow during the critical winter planting season, thus causing famine in Pakistan.

Since 1999, when India unveiled its plans, Pakistan has been demanding that the dispute go to arbitration. So far India has refused. Instead it appears to be rushing to complete the dam. The two countries are still talking, but the Indians are still building. And nobody can forget that since the last dispute over these waters, both sides have developed nuclear weapons. Could this beautiful spot be the flashpoint for the unthinkable?

The stakes are huge. The Indus and its tributaries make up one of the world's largest river systems. Most of the water rises in the rain-soaked mountains of the Indian Himalayas and the Hindu Kush in eastern Afghanistan before flowing through Kashmir and reaching arid Pakistan. For almost four decades the Indus Waters Treaty has been held up as a model for other countries with water disputes. But in truth it has been enforced largely at the insistence of the World Bank, which refused to give either country loans for new dams unless it signed up.

Even without the Baglihar imbroglio, tensions over the treaty have been growing. In 2002, during the most recent standoff between India and Pakistan over Kashmir, some nationalist politicians and commentators in India were so angered by terrorist incursions that they talked of unilaterally taking back the Indus tributaries currently allotted to Pakistan. Most vociferous were the Kashmiri politicians, who want the tributaries that flow through their territory for their own people. Jasjit Singh, the former director of the Institute of Defense Studies and Analyses in New Delhi, threatened, "Under normal circumstances, India would have continued to fulfill its moral obligation of sharing its water with Pakistan. But unusual circumstances call for unusual action."

Pakistan finds itself in the classic position of the downstream state—at the mercy of others for its most basic resource. But from the Indian viewpoint, things look quite different. As Indians see it, Pakistan got an extraordinarily good deal with the Indus treaty. Both Afghanistan and India contribute far more water to the Indus system than Pakistan does but take far less. It is not their fault if Pakistan has hitched its wagon so firmly to the river. Moreover, they grumble, Pakistan uses its share of the river with spectacular inefficiency while suffering no penalty. Having got itself into a position of undue dependence on the river, the Indians suggest, Pakistan would be better off not raising the specter of a water war—a war from which it could only lose in a manner devastating for its huge and fast-growing population.

The whole world should care about this obscure hydrological battle in a part of the world where few dare to go. Though tensions eased after Indian elections in 2004, the rumblings of the modern world's second water war never stop.

There are important reasons that warmongers with serious designs on power may hold back when it comes to dams. During two decades of guerrilla warfare in northern Mozambique in the 1970s and 1980s, fighters on both sides spared the giant Cahora Bassa Dam on the Zambezi. The reason? They expected to win—at which point they would need the dam. But in the twenty-first century, many terrorists have little desire to take power and few prospects of doing so. Therefore, they have less reason to hold back. And they will know that there are few better ways of inflicting terror than threatening to attack a dam, and few better ways of causing chaos than carrying out the threat.

In September 2004, the Chinese military suddenly sent helicopters, patrol boats, armored vehicles, and bomb-disposal robots to the site of its giant new dam at Three Gorges on the Yangtze River. Nobody is sure what was in the minds of the authorities, but the action followed press stories about the risk of terrorists hijacking a large ship on the Yangtze, packing it full of explosives, and ramming it into the dam. And it came only a few weeks after the Pentagon had reported that the Taiwanese military was considering just such an attack on the dam.

Three Gorges is the world's largest hydroelectric dam, holding back a reservoir 300 miles long that will before long contain 32 million acre-feet of water. Unleashing that water downstream would stand a good chance of creating the worst manmade disaster in history. It could dwarf the blasting of the Huayuankou dyke, which released the full fury of the Yellow River in northern China back in 1938. Even a credible threat of such an action would cause panic among the 350 million people downstream on the Yangtze. If the reservoir were emptied to neutralize such a threat, China's war machine would be crippled by the loss of electricity.

When the antidam campaigner Dai Qing interviewed scientists opposed to the construction of the Three Gorges Dam back in the 1980s, one military expert told her, "War is the key fact that determines whether or not we should construct the Three Gorges project. Are we going to make a sword of Damocles that will hang over the heads of future generations for decades to come?" They did.

Losing the West

It is a truism that water won the West. The 1450-mile Colorado, which drains a twelfth of the continental United States, is the lifeblood of seven states, delivering its water to burgeoning cities, feeding irrigation projects, and generating hydroelectricity. Since the 1930s, many of its beautiful canyons have been flooded to make reservoirs. So much water is captured that the amount that makes it to the sea has fallen to nearly zero, leaving the Colorado delta to shrivel in the sun. A once rich landscape where jaguars and beavers roamed, it has not seen fresh river water since 1993. And no river water means no silt to maintain the delta, which is growing ever more vulnerable to tidal erosion.

Nature has been sacrificed to the demands of Uncle Sam's farmers. But now the river itself is faltering. And from the snow-covered mountains of Wyoming and Colorado to the desert cities of California and Arizona, the beneficiaries of the Colorado are getting worried.

Two giant reservoirs control the flow of the middle reaches of the Colorado and insure supply to the lower states. The first, Lake Mead, was filled in the 1930s behind the Hoover Dam in Boulder Canyon. Then, in 1964, the Glen Canyon Dam drowned a series of spectacular gorges to create Lake Powell, which was named after John Wesley Powell, a one-armed Civil War hero who in 1869 made the first boat journey by a white man down the river.

The two reservoirs collect water when snowmelt in the Rocky Mountains fills the river and distribute it to cities and fields during the long summer growing season. Having more than four times the capacity of the river's average annual flow, they can also even out fluctuations between wet and dry years.

While urban areas are taking an increasing amount, most of the water abstracted from the river still goes for irrigating some 4 million acres of fields in the river valley and in Arizona and California. America has always subsidized farming in the West, and today perhaps $1 billion a year is poured into keeping farmers irrigating crops that they would not otherwise grow. And subsidies encourage waste. Every year several million acre-feet of water evaporate from reservoirs, farm ponds, and flooded fields, while much of what does get to crop roots is used to grow low-value crops such as alfalfa. Even in dry years, the presumption is that a wet year will be along soon. But that presumption looks increasingly foolhardy.

The Colorado is both legally and hydrologically one of the most regulated rivers in the world. But it is becoming clear that the legal and the hydrological

no longer mesh. A century ago, more than 20 million acre-feet of water flowed unimpeded to the Gulf of California every year. When lawyers shared out the river's waters between the states in 1922, on the eve of the dam-building era, they gave 7.5 million acre-feet to the upper basin states of Colorado, Utah, Wyoming, and New Mexico and another 7.5 million acre-feet to the downstream states of California, Arizona, and Nevada. With another 1.5 million acre-feet assigned to Mexico, that added up to around 16.5 million acre-feet.

That should have left water to spare, but ever since, flows have been diminishing. Since the compact was signed, the average flow has been 13 million acre-feet. From 1999 to 2003, the average sank to 7 million acre-feet—worse by far even than the Dust Bowl years of the 1930s. In 2002, it fell to just 3 million acre-feet. The U.S. Geological Survey says the Colorado hasn't seen a drought like this in five hundred years. Whether the cause is cyclical or global warming is unclear. But the survey says it can see no end in sight. The wet years that refilled the reservoirs have simply vanished.

Lake Powell was full to the brim in 1999, but by the end of 2004, after several years of drawing down the reserves, it stood three fifths empty. Its 170-mile-long reservoir could, if the drought continues, be empty by 2007. Lake Mead is scarcely better off. It is, of course, quite possible that between the time this is being written and the time you are reading it, the drought will have broken. Wet years can be very wet. The reservoirs could be full and the crisis could be over. Better rains in early 2005 brought some relief, but the crisis continued and the Bureau of Reclamation said that at average river flows, it would take the reservoirs ten years to refill. If the drought persists, the whole region is in deep trouble. And even if it breaks, most climate forecasts suggest that it will soon return with even greater force.

So in 2005 the states were starting to prepare for the worst. Since there was no more water, they deployed lawyers instead. The lower states, their attorneys contend, have a guaranteed entitlement, regardless of the state of the river: their share has to be provided from the reservoirs. If Lake Powell falls much lower, then the upper states will have to give up their share, emptying their own small reservoirs and halting abstractions from the river to meet downstream commitments. The upstream states, understandably, disagree. Colorado is not happy at being required to export three quarters of the snowmelt from its mountains when its own farms and cities are running low. Denver and Colorado Springs don't quite see why they have to shut off their sprinklers so Phoenix, Las Vegas, and Los Angeles can keep theirs on.

Most observers of the hydropolitics of the Colorado believe that the days when most of the waters of the mighty river could be used to irrigate crops cannot last much longer. Cities are demanding a larger share of the pie. And the size of the pie is declining. But if water shortages don't put the farmers out of business, salt may be the apocalypse awaiting the great American agrarian civilization. Salt could kill off parts of the United States as certainly as it did Ur and Urek. As Arthur Pillsbury, the doyen of water resources in the American West, said to me before his death, "The Colorado basin will eventually become salt-encrusted and barren because of salt." The only question is when.

All down the river, more and more salt is clogging up the system. It is flowing downstream from the headwaters in the Rockies. It is also being washed from soils and bedrock in irrigated areas like Paradox Valley in Colorado and Wellton-Mohawk in Arizona. The river and the extensive manmade irrigation and drainage networks that circulate its waters have also become a vast system for collecting and distributing salt. Each year about 10 million tons of salt enter the system, but virtually none reaches the ocean.

Almost all the water flowing down the Colorado leaves the river several times to irrigate fields and returns via drains. At each step it both loses volume, through evaporation, and picks up salt. So the concentration of salt in the water increases as it travels downstream. At its headwaters, the Colorado contains about fifty parts per million of salt. By the time it reaches the last dam, near Las Vegas, it contains more than 700 parts per million. Tens of millions of dollars are spent on farms every year trying to minimize the problem, but even so, annual crop losses from salt are currently estimated at $330 million.

Taking the Water to the People

It is the world's largest civil engineering project and aims to remake the natural hydrology of the world's most heavily populated nation. But it began rather inauspiciously, one morning in Beijing in April 2003, when the city's vice mayor, Niu Youcheng, accepted a bottle from a visiting local official. The bottle was filled with water from a reservoir on the Yangtze River 800 miles away in the south of the country. Its handover signaled the start of China's south-to-north transfer project, which is intended to be the ultimate solution to the desiccation of the Yellow River and the North China plain.

Some people see the scheme as an exercise in engineering hubris and a disaster in the making. But Chinese leaders say it is a logical extension of the grand schemes of great societies down the ages to remake their hydrology. Ancient Mesopotamia and Egypt both harnessed their rivers to feed their people. The Romans were famous for their aqueducts. Persian empires were built around laboriously excavated tunnels that delivered water from deep underground. But even at their greatest extent, these ancient works were mild modifications to natural drainage patterns. Today's engineers have bigger ambitions, diverting entire rivers onto distant plains. And China's scheme is the most ambitious ever to have got under way. It will, we are told, enable the Middle Kingdom to continue feeding itself as it has done for thousands of years. Maybe.

The south-to-north scheme will divert part of the flow of the Yangtze, the world's fourth biggest river, to replenish the dried-up Yellow River and the tens of millions of people in megacities that rely on it. The price tag is $60 billion, more than twice the cost of even Colonel Qaddafi's fantasy-world Great Manmade River Project. And it aims to deliver twenty times more water than Qaddafi's pipe dream.

Some of the water will come quite soon. In Beijing, people should be drinking Yangtze water regularly in 2007. Certainly it will be there in time to fill the swimming pools and pretty the streets with fountains as Beijing hosts the Olympic Games in 2008. Some of the water will take longer to get there, but within twenty years, say the planners, the project should annually be siphoning north three times as much water as England consumes in a year. The costs may be colossal, but China says the south-to-north project cannot be allowed to fail.

The project is actually three separate diversions. Two of them are already under construction. The first one will enlarge the existing Danjiangkou reservoir on the Han River, a major tributary of the Yangtze, and take its water north. The reservoir is already Asia's widest artificial expanse of water; enlargement will flood another 140 square miles and displace a quarter of a million more people. The canal north will be 200 feet wide and as long as France. As it crosses China's crowded plains, it will span 500 roads and 120 rail line and tunnel beneath the Yellow River through a giant inverted siphon.

The second will take water from near the Yangtze's mouth across Shandong Province on the North China plain and deliver it to the megacity of Tianjin, which has suffered chronic water shortages since the 1990s. Part of it will use the 2500-year-old Grand Canal, which was the world's largest artificial river in preindustrial times and the first to have lock gates. Today it is a sump for effluent from China's burgeoning industry, but there are plans to clean it up.

The third, western route is the most ambitious. It alone is expected to cost $36 billion. It will take water from the Yangtze headwaters amid the glaciera of Tibet and push it through tunnels up to 65 miles long into the headwaters of the Yellow River. There is no firm route yet, but several tributaries will be tapped, and there is talk of building the world's highest dam. This route will be the only one to deliver water directly into the Yellow River. The other two are intended to relieve pressure on the river by supplying water for cities and farms on the North China plain that currently take 8 million acre-feet a year from the river. But altogether the plan is to send some 36 million acre-feet of water a year from the Yangtze to northern China.

China's leaders love huge projects. Modernism lives on in their souls. With the World Bank claiming that China has already lost $14 billion in industrial production from water shortages, the scheme seems to them like a sound investment. But even as the first earth was dug, fears were growing about escalating costs. And academics and water planners I met in Beijing in 2004 raised a range of concerns.

The middle route, they said, could cause an ecological crisis on the Han River, taking a third of its water and worsening an already serious pollution problem. Wuhan City, a busy river port with a population of 3 million, could become a cesspit overnight. What, they wondered, about the cost of relocating refugees from the Danjiangkou reservoir? Could the filthy and decrepit Grand Canal really be cleared of pollution? Is the engineering intended for Tibet more than a figment of someone's imagination? And since China is trying to move to more realistic pricing for water, won't the transferred water be far too expensive for the intended recipients to buy?

China has a vibrant antidam community, which honed its arguments over the Three Gorges project. It sees the scheme as another megalomaniacal folly and wants the billions to be spent instead on improving Chinese water efficiency. Ma Jun, a journalist and campaigner in the mold of Dai Qing, says, "Chinese factories use ten times more water than most developed countries to produce the same products. Chinese irrigation uses twice as much." Even old-fashioned Chinese toilets use much more water than their Western counterparts. The United States has been able to grow its economy for the past thirty years without increasing water use, he says, and so can China. And in the long run, he argues, it would be easier to shift the focus of Chinese food production from the northern plains to the south—where the water is.

Sewage on Tap

Should dirty water be recycled? Nobody should countenance using neat industrial effluent to water crops. But sewage may be another matter. After all, even in Europe the concentrated muck created at sewage treatment plants is sometimes packaged up and sold as fertilizer for farmers. And in high-tech Singapore, officials announced plans in 2003 to add volume to the city-state's main drinking-water reservoir by topping it up with a 2.5 percent dose of recycled sewage effluent. That is small potatoes compared with London, whose inhabitants drink water that has been drunk and excreted several times as it makes its way down the Thames, being extracted and returned by towns such as Swindon, Reading, and Maidenhead before it reaches the capital. The advanced water-treatment technologies in use at every step make it as safe as water anywhere, despite its unappetizing recent history. The water-treatment

plants have become just another loop in the water cycle.

But across the poor, developing world, raw sewage is increasingly being used for irrigating crops. Chris Scott, of the International Water Management Institute, has conducted the first global survey of the practice. He has come to the staggering conclusion that perhaps a tenth of all the world's irrigated crops—everything from rice and wheat to lettuces, tomatoes, mangoes, and coconuts—are watered by the smelly, lumpy stuff coming out of the end of sewer pipes that empty the drains of big cities. Without it, much of the world would go hungry. In many countries—India, China, and Pakistan, to name just three of the biggest—there is very little sewage treatment, and yet a great deal of the sewage ends up being poured onto fields anyway, complete with disease-causing pathogens and sometimes laced with toxic waste from industry.

Scott estimates that more than 50 million acres of the world's farms are irrigated with sewage. And business is booming. The practice is most frequent on the fringes of the developing world's great cities, where clean water can be in desperately short supply in the dry season while sewer pipes keep gushing their contents onto the nearest open land all year round. In Hyderabad, the Indian city where he works, "pretty much 100 percent of the crops grown around the city rely on sewage. There is no other water available."

And however much consumers may squirm, farmers like it that way—first because the sewage is rich in nitrates and phosphates that fertilize their crops free of charge, and second because the supply is often much more reliable than clean water from rivers or irrigation canals, which means farmers can grow high-value crops that need constant watering, such as vegetables.

For these reasons, farms hooked up to sewage pipes make bigger profits than their rivals who rely on clean irrigation water. In Pakistan, farmers using sewage for irrigation typically earn $300 to $600 more annually than those without the benefit of sewage, says Scott. In West Africa, he met one farmer who grew twelve crops of lettuce a year on his sewage farm. You can see the benefits in land prices as well. In parts of Pakistan, it costs twice as much to buy fields watered by sewage pipes as neighboring fields irrigated with clean water. People downstream like the farmers, too. They are essentially operating a free municipal wastewater treatment service that stops rivers and reservoirs from stinking so much. And so, often secretly, do governments. The system feeds the people. In Pakistan, sewage irrigates a quarter of the country's vegetables.

Many would say that the risks outweigh the benefits. Those risks include disease among farmers and customers and environmental problems such as the buildup of heavy metals and unwanted nutrients in the soil and underground water reserves. "Right now," says Scott, "wastewater irrigation is in an institutional no man's land. Water, health, and agriculture ministries in many countries ban the practice but refuse to recognize that it is widespread." But he says, instead of trying to outlaw it or pretending that it does not exist, governments ought to regulate it. "We need to recognize that sewage is a valuable resource that grows huge amounts of food. So instead we should help the millions of farmers involved to do it better."

That means keeping sewage effluent for irrigating nonfood crops or crops that will be processed or cooked before being eaten. Obviously, it is more dangerous to pour sewage onto a field of lettuce than a field of cotton, or even sugarcane. And ultimately it means working toward treatment to make sewage safe. In a handful of countries—most notably Israel, Jordan, Tunisia, and Mexico—that is already happening. Sewage is treated to remove pathogens before being released to farmers. In these countries, recycling forms part of a national strategy for maximizing water use and making sure valuable nutrients are not wasted.

Mexico recycles enough treated wastewater to irrigate around 600,000 acres. I visited a giant new state-of-the-art sewage treatment plant at Juarez, El Paso's twin city on the Rio Grande, which treats half the city's sewage and delivers enough effluent down a canal to irrigate 75,000 acres. It is virtually the only source of water for crops downstream on the Mexican side of the river. Israel converts around 70 percent of the wastewater from its cities into treated effluent for irrigating export crops such as tomatoes and oranges. This is an effective addition to its national water supply of about a fifth.

This makes good sense. In urban areas, almost every drop of water brought to the city to fill faucets eventually leaves again as sewage effluent or in industrial waste. Where it can be made safe, it should not be wasted.

Out of Thin Air

Nature is a good fog-catcher … and may have some tips for would-be fog harvesters. On El Hierro, [one of the] Canary Island[s], people harvested fog droplets from the leaves of trees until a hundred years ago. Perhaps this practice was the origin of a report by Pliny the Elder, in Roman times, of a Holy Fountain Tree growing on the island. In Namibia, meanwhile, British zoologists recently discovered a beetle in the desert that has evolved a bobbled upper surface to its body with a pattern that is supremely efficient at capturing moisture from passing fogs. The hexagonal pattern of tiny peaks and troughs appears to push tiny droplets together to form larger droplets, which then roll off the beetle's back and into its mouth. The scientists, headed by Andrew Parker, of the University of Oxford, rigged up a prototype fog-catching surface based on the beetle's design, which captured five times more water than

Schemenauer's netting. So they patented it, and several companies are vying to make fog-collecting devices that can be put on the roofs of buildings, or even tents.

Is desalination of seawater the answer to the world's water woes? Some say so. Distilling seawater by boiling it and collecting the water vapor is an age-old activity. But modern distillation technology was developed by the U.S. Navy for operations on remote Pacific islands during the Second World War. Following that, large-scale distillation for public supply took off in the water-poor Gulf states, where they have plenty of oil to provide the necessary energy.

Today global desalination capacity is about 8 million acre-feet a day—roughly 3 percent of the global tapwater supply, though only a tenth of 1 percent of total water use. Most of the global capacity is still in the Gulf states. Saudi Arabia alone accounts for one tenth of world output, and in 2004 announced plans for six more plants costing a total of $5 billion. Islands where summer tourists have overwhelmed local water supplies have until recently made up most of the rest. In Mediterranean Europe, Malta gets two thirds of its drinking water from desalination. Greek islands like Mykonos have been doing the same for years, as have Caribbean islands such as Bermuda, the Caymans, Antigua, and the Virgin Islands.

Four fifths of the world's total desalination capacity still uses distillation, but since the 1970s, an alternative technology has grown in popularity. Reverse osmosis forces water repeatedly through a membrane that filters out the larger salt molecules and lets clean water through. Both technologies require large amounts of energy, whether to evaporate the salty water or to force it through the filters. Until recently, it cost several dollars to produce 265 gallons of unsalty water—typically a hundred times more than conventional water supplies. But costs for reverse osmosis have come down, and more cities are buying into the technology.

Tampa Bay, Florida, and Santa Cruz, California, have both taken the plunge, and more reverse-osmosis plants are slated for Houston, Texas; Cape Town, South Africa; and Perth, Australia. In Spain a new government elected in 2004 swiftly abandoned its predecessor's plans to pump water cross-country from the wet north to relieve the arid south, in favor of twenty reverse-osmosis plants. They are expected to meet slightly over 1 percent of Spain's total water needs.

The cheapest desalinated seawater is now in Israel, where one of the world's largest reverse-osmosis plants has been built on the Mediterranean coast at Ashkelon. Israeli water economics are notoriously opaque, but the government claims to be able to deliver water at around fifty U.S. cents per 265 gallons, around a third of the production cost in Saudi Arabia. (More pertinently for Israelis, it compares with the thirty cents it costs to pump freshwater from the Sea of Galilee to Tel Aviv and the two dollars to buy water by the tankerload from Turkey.)

Such prices are encouraging cities in less extreme circumstances, and in cooler and wetter climates, to join the reverse-osmosis revolution. During 2004, China announced plans for a giant plant for Tianjin, the country's third largest city, where water shortages have been endemic for years. Even more surprising, Britain's Thames Water announced that it will build a $400 million reverse-osmosis plant on the Thames estuary in east London to process water during droughts. It will be able to produce up to 120 acre-feet a day, enough to meet the domestic needs of almost a million people.

The boom in desalination is beginning to alarm environmentalists. One problem is what to do with the salt extracted from the seawater during the process. It emerges as a vast stream of concentrated brine. Most plants, naturally enough, dump it back into the sea. But this salty wastewater also contains the products of corrosion during the desalination process, as well as chemicals added to reduce both the corrosion and the buildup of scale in the plants.

Maybe this pollution can be fixed technically one day. But what can't be fixed is the huge energy demand of desalination. A typical reverse-osmosis plant consumes six kilowatt-hours of electricity for every 265 gallons of water it produces. Most of the power, inevitably, comes from burning coal, oil, and other fossil fuels. So while desalination could conceivably become a viable source of drinking water in coastal regions around the world in the coming decades, it would be at the expense of an extra push toward climate change.

It is also hard to see desalination ever penetrating the agricultural market, where the majority of the world's water is currently used. At the end of the day, desalination seems like an expensive high-tech solution to a global water problem that is overwhelmingly caused by wasteful use. Like enormous engineering projects for shifting water around the planet, it is a supply-side solution to a demand-side problem.

Catch the Rain

In China, it was Chairman Mao and his Cultural Revolution that began the revival of the ancient tradition of rainwater harvesting. In India, it has been a mixture of swamis and scientists, schoolteachers and even policemen. Haradevsinh Hadeja is a retired Indian police officer. He loves playing cricket for his village team. When we met, he had a broken arm caused by a fast bowler from the next village. But win or lose at cricket, his village of Rajsamadhiya, in the backwoods of Gujarat, always excels at water.

His fellow villagers say Hadeja is a near-magical diviner of water, and you can see why. Around the village, he has transformed a desertlike landscape of desiccated fields and empty wells into a verdant scene of trees, ponds, full wells, and abundant crops. With no piped water, most of the other villages in the area rely on tankers to provide drinking water for much of the year. They have little left to irrigate their crops. But here, says Hadeja, "We haven't had a water tanker come to the village for more than ten years. We don't need them."

He worked this miracle by catching the monsoon rains. Not, like the Chinese, in purpose-built cellars and cisterns, but in ponds. And the villagers don't use the water directly from the ponds. They allow it to percolate into the soil to refill underground water reserves and replenish their wells. "There is no more rain than before. We just use it better. We don't let it wash away," Hadeja says. The village has twice as much water as before, and wells find water at only 20 feet down, whereas once the water had to be hauled up from more than 100 feet. The contrast between Rajsamadhiya and surrounding areas where water tables are falling is extraordinary.

I went to look. The heart of the village appeared conventional enough: a gaggle of single-story houses leading from a small square out toward the fields. But on the paths there were thousands of fruit trees, where most villages are treeless. Under their shade were piles of mangoes and watermelons. And out among the small fields growing wheat and vegetables and groundnuts, there were the ponds—lots of them. "We have forty-five water-collecting structures altogether," said Hadeja as we walked past a line of women washing clothes in a pond. "This one gets its water from land up to 3 miles away," he said.

The ponds are arranged along the routes that the monsoon water takes as it drains through the village. Rather than trying to get rid of the floodwater, Hadeja has redesigned the village's drainage to slow the water's passage long enough for it to collect in specially dug ponds. The water passes from one pond to the next in a slow cascade, seeping through the soil to refill the aquifer all the way.

Hadeja's second innovation has been to manage how people use the water. As a former policeman he has some authority, and though all decisions are taken by the village council, his word has obvious force here. (In fact, he told me quietly, the police and other authorities never come to the village now. They leave it all up to him.) Under Hadeja's law, nobody is allowed to take water directly from the ponds, and farmers are banned from growing the thirstiest crops, like sugarcane. "There is no point in catching more water if we only waste it," he said.

News about this remarkable village has spread around India and beyond. One foreign scientist brought satellite images of the village that showed hidden cracks in the geology through which water was flowing. Hadeja slowed the flow by plugging the cracks with concrete. But mostly, as in Gansu, water scientists are coming here to learn about rainwater harvesting, though the water diviner insists that there is little to learn. "I am an uneducated person," he said. "I saw that people were leaving the village and I wanted them to stay. That meant finding more water. So I tried to catch the rain."

And in truth, this is no miracle. Hadeja has tapped into an old tradition and developed it. In India, you can still see abandoned ponds and lakes dotted across the countryside and on wasteland in cities. Until the early nineteenth century, much of India was irrigated from shallow mud-walled reservoirs in valley bottoms, which captured the monsoon rains each summer. The Indians called them *tanka*, a word the English adopted into their own language as "tanks."

Most of the tanks were quite small, covering a couple of acres at most, and irrigating perhaps fifty acres. Farmers scooped the water from the tanks, diverted it down channels onto fields, or left it to sink into the soil and refill their wells. The tanks served other functions, too. Some were stocked with fish, All were prized for the silt brought into them by the rainwater. Farmers guarded the slimy, nutrient-rich mud in their tanks almost as much as the water. They dug it out to put onto their land and turned silted-up former tanks into new farmland.

Overall, across India, researchers estimate there are around 140,000 tanks, either still in use or abandoned. Tamil Nadu has the most, approaching 40,000, covering several percent of the land surface of the state and still irrigating around 2.5 million acres. Karnataka estimates it has 35,000. Every region has its own design. In the Thar Desert of northwestern India, people channel runoff into manmade desert depressions called *khadin*, creating wet soil for planting wheat or chickpeas.

The system thrived until the British took charge in India. Though full of admiration for some of the grand Indian water structures on rivers, British water engineers largely ignored the village tanks, apparently not realizing that they were the way India fed itself. Tanks passed into a kind of forgotten underworld, used while they served a purpose but unrecognized by officialdom and rarely repaired or cleaned out. As the British and later the Indian government promoted more modern water-gathering technologies, they gradually fell into disuse. But today, as the formal irrigation systems established on the Western model fail across the country, and as farmers are having to pump from ever greater depths to retrieve underground water, the old tanks are starting to be restored.

All across India, groups are harvesting the rain either for direct use or to revive underground water reserves. Besides tanks and *khadin*, there are also check dams. These are barriers constructed in small streams and gullies to hold up the monsoon rain long enough for it to percolate underground. In Rajasthan, a government scientist, Rajendra Singh, gave up his job, taught himself traditional water-harvesting skills, and went out to the edge of the Thar Desert, where villages were dying for want of water. He encouraged the locals to install check dams. Now his movement has 4500 of them collecting water in several hundred villages, forty-five permanent staff members, and a grant from the Ford Foundation.

Whole landscapes are being transformed. In Limbadia, a village in western Gujarat, the water table was 500 feet down and falling fast until villagers built a series of check dams. Soon afterward several wells began spurting water at the surface. Tushaar Shah, at the International Water Management Institute in Gujarat, estimates that across Rajasthan, some 2500 square miles of land are being newly managed to capture the rains, "with dramatic impact on groundwater recharge and the revival of dried-up springs and rivulets." Water tables have risen so much in Rajasthan that five ancient desert rivers—the Ruparel, Arvari, Sarsa, Bhagani, and Jahajwali—have returned to the map.

In Karnataka, on the plains west of Bangalore, I visited a group called the BAIF Development Research Foundation, which has helped farmers dig 350 ponds across four valleys near the town Adihalli. Water flows from one pond to the next in a slow cascade. The result is more water in village wells, year-round farming of grains, and improved yields of cash crops like coconuts and chiles, cashews and mulberries, vegetables and rice. Incomes have doubled and sometimes quadrupled.

"There used to be water on the land ten days a year; now they have it all year round," said the director of the project, a Gandhi disciple named G. N. S. Reddy. One local farmer told me with delight, "I can irrigate my vegetables from the pond once a week, and afterward there is still water in the pond." As Reddy and I walked back to the road, with birds singing in the trees, we met women goatherds who had walked several miles to the ponds. "Where we live, there is no water. So we come here. They let us use the water," one told me.

I met a husband and wife who a few years before had given up farming and become migrant laborers for much of the year—pretty much the bottom of the heap, even in India—after their 325-foot-deep borehole went dry. But now, thanks to the local network of ponds, the water table was back to 130 feet and they were home again. When we dropped by, they were wielding a long hosepipe to irrigate rice and gherkins. Ratnama, the wife, said with a broad smile, "We used to steal fodder from other people's land for our animals. Now other people steal ours. We've built a new house and sunk another borehole. We are not begging anymore. We can even get loans."

Rainwater harvesting is becoming a widespread social movement, uniting many strands of Indian society. Pandurang Shastri Athaval, a Vedic scholar in western India known to his followers until his death in 2003 as Dada, preached a simple life, in which commonly owned resources such as water were revered and their protection was seen as an act of devotion. With this creed, Dada persuaded tens of thousands of villagers to construct low mud walls on their fields to divert the monsoon rains directly down their wells. Some 300,000 wells have so far been adapted to receive rainwater.

"Even cities can do it," says Sunita Narain, the director of the Delhi-based Center for Science and the Environment, an outspoken advocate. "In parts of Delhi where old tanks and ponds have been cleared of garbage and refilled with water, the water tables are rising." Delhi could, if it got organized, obtain a third of its water from harvesting the rains, she says. In Bangalore, in India's Silicon Valley, they are trying to boost the aquifers by rehabilitating the city's sixty ancient lakes. The city's water is among the most expensive in Asia, because to keep the city supplied, the authorities pump 400,000 acre-feet up 1600 feet from the distant and drying Cauvery River. The Bangalore architect S. Vishwanath says, "The city gets about 800,000 acre-feet of rainwater a year. We need to use that instead."

Vishwanath is part of a movement to revive the city's own water reserves called the Rainwater Club. I saw houses, offices, apartment complexes, schools, even a planetarium that were catching water from roofs, gardens, paved areas, and parks, thanks to its work. Vishwanath's own eco-house captures enough water from the roof during the monsoon season to flush toilets and run the washing machine for most of the year. An evening's rain can be enough to run the house for a week. From the roof beside the cistern, Vishwanath pointed down the street to a city water tower. "They supply water about every four days. I have water from my roof anytime I want it," he said.

But as heartening as these individual initiatives are, the roots of the Indian rainwater-harvesting revolution lie in the communal rehabilitation of aquifers. Shah set up his groundwater research center in Gujarat partly to learn the lessons of people like Hadeja, Singh, and Dada. He says the vital factor in their success is that the initiatives happen at the village scale. Few individual farmers can success-fully catch their own rain and store it underground; it would quickly dissipate into the wider aquifer. But when an entire village does it, the effects are often spectacular. Water tables rise, dried-up streams flow again, and with more water for irrigation, the productivity of fields is transformed.

Shah says the rainwater-harvesting movement is "mobilizing social energy on a scale and with an intensity that may be one of the most effective responses to an environmental challenge anywhere in the world." It is, he points out, completely autono-mous from government. "It emerged on its own, found its own source of energy and dynamism, and devised its own expansion plans."

By some estimates, 20,000 villages in India are harvesting their rains. Not much, perhaps, in a coun-try that boasts a million villages, but it is a start. And however the "social energy" is created—whether by the force of a personality like Hadeja's or through religious devotion or a Gandhian ethos—some kind of communal water ethic seems to be the magic ingredient. That and a belief whether expressed in religious or secular terms, that, as Dada put it, "If you quench Mother Earth's thirst, she will quench yours."

Learning to Love the Floods

[Europe's] summer rains are getting fiercer, owing possibly to global warming. Glacial melting adds to the flow. But something also has gone badly wrong with the way Europe manages its rivers. Engineering intended to prevent floods is, it increasingly seems, conspiring to create them. We've seen how the operation of large dams can cause floods. But this problem extends to the entire management of a river system. All the major rivers that flooded in 2002 had been engineered specifically to banish floods. Their wetlands had been drained and their meanders straightened. The rivers were held in check behind high levees, and all impediments to having the water rush to the sea had been dynamited or chain-sawed from their path. But instead of eliminating floods, all this effort simply sped the water to the nearest bottleneck, where the floods became concentrated.

The strategy of rushing floodwaters to the sea, Europe's administrators were forced to conclude, has failed. "Flood peaks are higher and more dam-aging in places where wetlands and floodplains have been cut off from rivers, channeling more water into an unnaturally small space," Günther Lutschinger, a floodplain ecologist for WWF Austria, told me as authorities cleared up the mess in 2002. At the same time, many of the continent's mountains have lost most of their forests, reducing their capacity to absorb heavy rains.

In a single day at the height of the floods of 2002, most of central Europe received more than 7.8 gal-lons of rainfall for every square foot of land. Most of it surged within hours into the region's rivers, where 80 percent of the rivers' floodplains had been barricaded off by 600 miles of dykes, and where "improvements" had shortened the rivers' main channels by a quarter. The resulting floods came downstream in a great rush. And when the dykes failed, Prague and Dresden were engulfed.

The European Union is trying to improve forecasting of intense rainfall and to model how such rains affect river flows better. That may help cities prepare evacuation plans, but it won't stop the floods. Europe's crowded floodplains are at constant risk of inundation. So is it time for a new approach—time to learn how to love the floodwaters, to embrace them instead of trying to get rid of them?

There are two ways of beating floods. You can eliminate the water fast, draining it off the land and rushing it to the sea down high-banked rivers that have been reengineered as high-performance drains. Or you can encourage nature to hold on to it, letting it go only when the rains have stopped and the rivers are lower. Until recently, engineers preferred Plan A, the fast option. But however big they dug city drains, however wide and straight they made the rivers, and however high they built the riverbanks to keep the water on its path, the floods kept coming back to taunt them. From the Mississippi to the Danube, the flood-free future has failed to arrive. Dykes turn out to be only as good as their weakest link—and nature will unerringly seek that out.

By trying to turn the complex hydrology of rivers into the simple mechanics of a water pipe, engineers have often created danger where they promised safety and intensified the floods they intended to prevent. So now, more and more engineers are turning to Plan B. They are holding back the floods, capturing the water in fields and tearing down the banks and dykes and levees to give rivers back their floodplains. They are putting back the meanders and marshes to slow down the flow. They are even plugging up the drains on farms and in cities, encouraging the floodwaters to percolate underground. Rivers need room to flood, they say. And cities need to become more porous.

"The recent floods have provoked a completely new way of thinking," says the hydrologist Piet Nienhuis, of the University of Nijmegen in the Netherlands, a country that knows a thing or two about floods. "Rivers have to be allowed to take more space. They have to be turned from flood chutes into flood foilers."

Some call this soft engineering, going with the flow of nature. England's Environment Agency, which was given an extra $250 million a year to spend after floods in 2000 that cost $1.5 billion, is an advocate. It says, "The focus is now on working with the forces of nature. Towering concrete walls are out, and new wetlands are in." The soft engineers want to go back to the days when rivers took a more tortuous path to the sea and floodwaters lost impetus and volume while meandering across floodplains and idling through wetlands and inland deltas.

Modern cities could hardly be better constructed to create floods. They are concreted and paved and asphalted and culverted so that rains flow quickly into rivers. But the new breed of soft engineers wants our cities to become porous, so that they can capture and store that water instead. In Germany, Berlin is leading the way. The city's massive redevelopment since the fall of the Berlin Wall has been governed by tough new rules to prevent its drains from becoming overloaded after heavy rains. Harald Kraft, a city architect specializing in the new systems, says, "We now see rainwater as a resource to be kept rather than got rid of at great cost." Here is genuinely radical thinking. New ideas for flood protection begin to connect up to new ideas for harvesting the rain.

Take the giant Potsdamer Platz, a huge commercial redevelopment in the heart of Berlin. The city council has set a maximum limit for drainage from the site of just 1 percent of the potential runoff during a big storm. If the project doesn't meet the target, the drains will back up. Simple as that. So architects have designed the buildings to capture rainwater from the roofs. The water will flush toilets and irrigate roof gardens. Meanwhile, water falling to the ground will go to fill an artificial lake or percolate underground through porous paving. All told, the high-tech urban development can store a sixth of its annual rainfall and reuse most of the rest. It needs fewer drains. And it also needs fewer water-supply pipes.

New housing developments across Berlin are adopting similar technology. In the Zehlendorf suburb, rain from the roofs, gardens, and drives of 160 houses is collected to irrigate parkland. Not a drop of water leaves the area. Harzahn, a drain-free development of 1800 homes packed onto just 75 acres, features roads built of cobblestones to allow rainwater to percolate through the gaps into the soil beneath. Could this be done on a citywide scale? The test case could be Los Angeles. This is where the benefits of combining modern thinking on flood protection and water supply could really pay big dividends.

LA, as the song says, is a great big freeway. With hard, impervious surfaces covering 70 percent of one of the world's largest cities, drainage is a huge task. LA has spent billions of dollars digging drains and concreting riverbeds to take the water from occasional intense storms to the ocean. And still many communities regularly flood. The city engineers, locked in old thinking, want to spend $280 million raising the concrete walls on the Los Angeles River by another 6 feet. Meanwhile, with terrifying irony, the desert city is shipping in more water by pipe and canal from hundreds of miles away in northern California and the Colorado River in Arizona to fill its faucets and swimming pools and irrigate its golf courses. Is this sensible?

Like Phoenix, southern California has a self-image as a desert region that can revel in water. It has the biggest swimming pools, the highest fountains, and the most water-hungry agricultural crops. No problem. It prides itself on being able to pay. Marc Reisner titled his memorable book on the state's water politics *Cadillac Desert*, a phrase that sums up the mindset. But this strutting hides a serious misreading of the state's geography, which sounds like bad planning. "In LA we receive half the water we need in rainfall, and we throw it away. Then we spend hundreds of millions to import water," says Andy Lipkis, an LA environmentalist. "We should be catching our own rain before trying to buy other people's."

Lipkis, along with citizens' groups like Friends of the Los Angeles River and Unpaved LA, want to beat the urban flood hazard and fill the faucets at the same time by keeping the floodwater in the city. They call their dream the "porous city." They are working with schools and other places to convert asphalt areas into soft ground to soak up rain. And they believe it is a realistic proposition to do this across the city.

It could happen. The city authorities have recently established a watershed management division, and in 2004 they launched a $100 million project to road-test the porous city in one poor flood-hit community in Sun Valley. The plan is to catch the rains that fall on thousands of driveways and parking lots and rooftops in the valley before it gets to the drains. Trees will soak up water from parking lots. Homes and public buildings will capture roof water for irrigating gardens and parks. And road drains will pour water into old gravel pits and other leaky places that should recharge the city's underground water reserves. It's a direct replacement for a planned storm drain. The result, the city hopes, will be less flooding and more water for its inhabitants.

Plan B holds that every city should be porous and every river should have room to flood naturally. It sounds expensive and Utopian, until you realize how much we already spend trying to drain our cities and protect our floodplains—and how bad we are at doing it. Katrina only underlined that. Plan B, going with the flow of the river, really is the better way. The question now is whether that can form the basis for a wider philosophy for managing rivers and the water cycle.

More Crop Per Drop

Can Pepsee save the world? Pepsees are small tubes made of light, disposable plastic and designed to encase individual ice candies—lollipops, or whatever you choose to call them. They are manufactured all over the world and are just about the most disposable product imaginable. In India they are made under the Pepsee brand name. Millions of roadside vendors across the land buy the polyethylene tubes, which come in long rolls that can be torn off at perforations stamped into the plastic every 8 inches or so.

Sometime around 1998, somewhere in the Maikal hills of central India, someone—perhaps a farmer with a sideline of selling ices—started using Pepsee rolls for another purpose: to irrigate the fields. The farmer had discovered that the rolls of plastic tubing made perfect cheap conduits for distributing water to plants. Shilp Verma, of the International Water Management Institute in India, says, "It is not very clear how and exactly where the innovation first started, but it spread among farmers like wildfire." The farmers took the rolls, laid them down rows of crop plants as close as possible to the roots, and poured water in at one end. As the water ran down the tubes, it dripped through the perforations to water the plants.

At the start, says Verma, there was one problem. Algae grew in the wet, sunny environment of the tubes. But Indian innovation was equal to the task. Wise to the new use for their product, the manufacturers produced an even cheaper version made of black recycled plastic and known to farmers as the Black Pepsee. With light excluded, the algae problem was solved. This startlingly successful exercise in lateral thinking has produced the first dirt-cheap method of providing drip irrigation for poor farmers.

Farmers across the world have traditionally irrigated their fields by pouring on water indiscriminately. Some reaches crop roots; most does not. The result has often been waterlogged soils, a buildup of salt, and growing water shortages. At a conservative estimate, two thirds of the water sent down irrigation canals never reaches the plants it is intended for. Technologists early in the twentieth century got to work on the problem. First they came up with the sprinkler, which doused fields in a fine spray of water from a central pivot. That saved the worst excesses of overirrigation but was equally indiscriminate and lost a lot of water to evaporation. It was also expensive and required energy to spray the water around.

The next advance was drip irrigation. Various people claim this idea for themselves, but credit is normally given to an Israeli engineer called Symcha Blass, who retired to the Negev Desert in the early 1960s and, as he told it, one day noticed how a large tree grew in the desert because it was right next to a slowly dripping faucet. He mused on the matter and subsequently invented a narrow tube that could deliver water under pressure and drip it close to the roots of plants. He filed a patent in Tel Aviv in 1969. The trick, as much as anything, was timing. His idea coincided with the development of plastics that made such a system economical for the first time, and with the growing realization in arid countries like Israel that water was in increasingly short supply.

Drip irrigation can take many forms. Mostly it is high-tech, with water pumped down pipes under pressure and sent into side pipes from which sophisticated "drippers" deliver it to roots. Such systems can include flow meters, pressure gauges, and even soil-moisture gauges to optimize delivery and keep losses to a minimum. Today large farms in California, Tunisia, Israel, and Jordan specialize in such systems. In Jordan, drip irrigation has reduced water use on farms by a third while raising yields. Israeli farmers have raised water productivity fivefold in the past thirty years through a mixture of drip irrigation and the recycling of treated urban wastewater onto their fields.

The technology should have taken off. But it hasn't. It remains virtually ignored by the mass of small farmers in poor countries, who face some of the worst water shortages. In India, for instance, despite strong government promotion and subsidies, drips irrigate fewer than 1 percent of fields. "It is still largely seen as a technology for gentleman farmers," says Verma. This is perhaps not surprising, since the full kit can cost more than $800 for one acre, and even stripped-down systems developed for poor contries cost at least $200.

Another reason is that most farmers in most places at most times get their water at heavily subsidized prices—a tenth of the real cost is typical everywhere from India to Mexico and Pakistan to California. And when farmers pump water from beneath their fields, they pay no more than the cost of a pump and electricity, which itself is usually highly subsidized. Except in an immediate crisis, there is little incentive to save water. So what will turn the tide? More realistic pricing of water would certainly help. But so too would appropriate technology. And that means going back to basics.

Drip irrigation, it turns out, is not a brand-new technology. At least two thousand years ago, Chinese farmers were making small holes in earthenware pitchers and burying them in the soil. They would go around their fields every few hours to refill the pitchers, which then simply leaked water into the root zone. This indigenous system is also known in India and parts of Africa and the Middle East.

Indian farmers also have a tradition of irrigating from hollow bamboo tubes pierced with holes and laid out along fields. More recently they have adapted the bicycle inner tube, making a few extra punctures to allow a steady drip of water into the ground. But you can get your hands on only so many inner tubes, so now the Pepsee is spreading fast across India. No wonder it took off. A 2-pound roll of the stuff can be bought for fifty rupees (about one dollar). Even allowing for the necessary plumbing to deliver water, the overall cost—around 1000 rupees, or $23, an acre—is less than a tenth of that of even the cheapest conventional systems.

In the face of escalating water problems, agricultural researchers all over the world are adopting a new philosophy of "more crop per drop." It is a huge turnaround from the green-revolution days, when the philosophy might reasonably have been called "the more water, the better." Raj Gupta, the director of a large agricultural research center run by the International Maize and Wheat Improvement Center in the heart of Delhi, says, "For the first time, we are starting to measure crop yields in terms of the tonnage produced for a given amount of water, rather than a given amount of land."

Gupta holds that cheap drip-irrigation systems like the Pepsee could save India from running out of water. To go with it, he has a whole group of changes to current farming methods. They include planting crops on raised beds, reducing plowing, and leveling the land to prevent waterlogging. Together, he believes these approaches could reduce water use across northern India, the breadbasket of the country, by a third or more, and save millions of farmers in the south from penury as the groundwater bubble bursts. And what will work in India should work in China and Colombia, Mexico and Mali, Libya and Lesotho.

Rice is an early target for the "more crop per drop" crusade. The world's most popular grain consumes more water than any other food crop, typically twice as much as wheat for every ton of output. More than a third of all the water abstracted from rivers and groundwaters on the planet goes to irrigate rice paddies in Asia. But with half of the world's rice grown in water-stressed China and India, that level of production cannot continue.

Experiments in India and at the International Rice Research Institute in the Philippines show that rice can be grown with much less water. The trick is to abandon the traditional method of growing seedlings in nurseries and then transplanting them into paddies that have to be kept flooded. Much less water is needed if they are instead planted directly into muddy soil. Some varieties can already be grown this way. If those varieties were more widely adopted, they would cut water use by a fifth—and require less labor into the bargain. Other crops offer similar benefits from water-saving methods of cultivation.

Researchers say this new approach amounts to a "blue revolution" to follow the green revolution. The Washington-based International Food Policy Research Institute (IFPRI), which is part of a network that also includes the international water management and rice research institutes and Gupta's maize and wheat center, says it is vital that technologies like drip irrigation be more widely adopted. Together, they could bring cost-effective global savings in water use of more than a fifth.

Already governments in water-short regions are eager to take up the challenge. Centrally planned China is in the forefront. Unable to wait for Yangtze water from the south-to-north project, Shandong Province on the North China plain—where the people get whatever water is left in the Yellow River after every other province has had its fill—responded to water riots in 2000 by agreeing to spend $6 billion on water conservation on farms.

Meanwhile, there is plenty to be done at plant-breeding stations. Most modern high-yielding plant varieties are hopeless water guzzlers. Only now are breeders turning their attention to producing varieties that use water efficiently. New rice varieties, the researchers say, could one day cut water use by as much as half. But another step is required. There has to be a reconsideration of where crops are grown. We need to fit the crops grown to the availability of water rather than attempting to do things the other way around. Many parts of the arid world simply need to give up growing water-guzzling crops.

Cotton is perhaps the worst offender. The fiascos in Central Asia are the most obvious case in point, but most of the cotton-growing in the world is destroying rivers. The cotton growers of the Indus should be halted. Many researchers go further. Gupta says that much of northern India should stop growing rice altogether and switch to wheat and corn. Similarly, the dairies of Gujarat should be weaned off cattle fed on irrigated alfalfa in the state. In China, many argue that the country should shift its entire farming effort from the arid north, where the Yellow River is running out, to the wetter south. Better that, they say, than spending tens of billions of dollars to take the water north.

Mark Rosegrant, at IFPRI, says the effects of all this could be profound. Cutting abstractions for agricultural irrigation water could leave an estimated 800 million acre-feet of water in the world's rivers each year. "Many planned dams will be cancelled," he says. From the Rio Grande to the Yellow River, the Indus, and the Nile, dried-up rivers could resume their flows. A start could be made on reviving the Aral Sea.

Shadow Cities

A Billion Squatters, a New Urban World

By Robert Neuwirth

Introduction

I had the pleasure of meeting Robert Neuwirth when my dear friend, Paul Sneed, now on the Portuguese faculty at U. Kansas, brought him to SDSU in 2006. In *Shadow Cities*, Neuwirth describes his visits to four squatter cities, also called "slums" or "informal developments." In these four cities spread over four continents, Neuwirth lived with the people whose story he tells. He shared their experience as he interviewed them, and their tales serve to humanize an oft-overlooked aspect of the global population crisis. We often hear there are too many people in the world. Those of us who do a bit of reading know this situation is poised to get much worse before it gets better, as the population grows from its current seven billion to at least nine billion by around 2050. Often, however, we miss the fact that *almost all* of this increase of two billion in human population will a) come from the developing world and be poor; b) have migrated from rural poverty, drawn to the big city by (usually false) ideas of newfound prosperity; and c) will find themselves without the means to build, buy, or rent a legal dwelling. They will, in other words, be squatters. There are a billion of them already, and by 2050 there will be three billion: *One-third of the human race.*

How did this come to be? There are as many reasons as there are threads in the fabric of our global future. The Industrial Revolution is a huge factor. Before the ready abundance of fossil fuels to feed machines to do labor, most of the labor had to be done by humans (animals can do labor as well, but this is an inefficient process). The result was that it took nine farmers to raise ten people's worth of food. When oil became available to make tractor fuel, natural gas to make ammonia fertilizer and coal to make electricity for industrial machines, it took fewer farmers to produce the same amount of food. Some of them became redundant, what Vandana Shiva calls "disposable people"; they went to the city rather than starve.

This happened first in Europe and North America as these areas developed; now the rest of the world is following suit. From the northeast of Brazil, from western China, from the African savannas, they go to the cities. Drought, climate change, and the conscious "improvements" of the Green Revolution have exacerbated this trend. The rural-to-urban migration of 400 million Chinese peasants over the next 20 years has been called "the greatest human migration in history," but it is merely the largest of many such migrations, covering the whole planet.

Neuwirth has many words for those who would denigrate these people as lazy and the places they move to as "slums"; to these, I can only add *look in the mirror*. It is easy to forget that two generations ago Italians, Irish, European Jews, and former slaves filled the "slums" of our eastern cities. Yes, crime bred in the poverty and neglect of these areas; so did a great deal of manufacturing skill, many ethnic, cultural, and musical traditions that make America so wonderful, and so did a proud and steely union movement that stood up to the 100-hour workweeks, deadly fires and industrial accidents, and child labor of the time and demanded better conditions for working people. Neuwirth beautifully depicts how the urban underclass of the developing world is repeating and innovating on this process. I'll shut up now and let him talk.

Preface: Out of the Shadows

I have no past, and an excess of future.

—Martín Adán

Four of us edged into the room, our knees knocking against the small table that served as tea tray, desk and impromptu kitchen. Elocy Kagwiria Murungi scratched a match and lit a lantern and a small kerosene burner so she could cook us supper. The uncertain light bounced off the corrugated metal walls, casting wild shadows on our laces and on the meager belongings around us. "I act like lice, the way lice act," Elocy said as she rinsed a pot. "I burrow in and scrape out an existence."

Like most of her neighbors. Elocy came to Kibera, the largest mud hut neighborhood of Nairobi, Kenya, with nothing except a willingness to work. At first she thought Kibera was sickening, ghastly, terrible—an open sewer, not a community. But it was the only place she could afford. So she sold stationery on the street for a few shillings. She washed dishes at a local eatery for a few shillings more. Now she is a teacher in a school for street

children located right inside the squatter community. The pay is not great, but the work is challenging and important. Here in Kibera, Elocy has improved her life and opened up new horizons for herself and her young son Collins.

Yet this spirited, intelligent and hard-working woman describes herself as a parasite.

January 2005: Officials in Mumbai pushed forward with a brutal scheme to demolish squatter communities around the city. The goal, authorities said, was to transform Mumbai into Shanghai, to chase away the chaos of the shantytowns and produce a city open for development. The politicos didn't care where the 300,000 people they evicted would go—as long as it was outside the city limits.

May–June 2005: Zimbabwe followed Mumbai's lead, embarking on a ferocious anti-squatter putsch. Police and soldiers ejected 700,000 people from their informal abodes in Harare and Bulawayo, the country's two largest cities. President Robert Mugabe christened this program Operation Murambatsvina—Operation Drive Out Trash. And these citizens were treated like trash, some left to sleep in the open in the middle of the winter, others trucked to the countryside and dropped at the side of the road with a warning that if they returned to the cities they would be killed.

December 2005: The Nigerian government evicted thousands and smashed their homes in Abuja and Lagos. These were not shanties or shacks. These were concrete and brick houses that people had labored mightily to build. But their sin, according to authorities, was that they did not get proper planning approval. So the homes had to go and the people were forced onto the street.

Despite the hardships, newcomers are still seeking their future in the world's cities. Every year, close to 70 million people leave their rural homes and head for the cities. That's around 1.4 million people a week, 200,000 a day, 8,000 an hour, 130 every minute. And the migration does not seem likely to stop, By 2030, there will be 2 billion squatters. And, by the mid-point of this century, there will be 3 billion squatters—more than 1/3 of the people on the globe.

If the world's cities want to keep pace with the influx, the United Nations says, they must build 35 million homes a year. That's 96,150 a day, 4,000 homes an hour, 66 homes a minute, one every second. And this would only maintain the equilibrium. It would not house the billion who are living as squatters today.

Even the UN doesn't believe this can happen. So it has proposed a stripped down approach: to prevent the formation of huge new squatter communities all over the globe, the world's cities should build homes for 670 million people over the coming 15 years. This, the UK estimates, would cost $294 billion. A mammoth amount, to be sure. But some simple math brings the number down to earth. We could raise that sum by collecting $3 a year from every person on the planet. Even Elocy and my other squatter friends might be able to swing that.

But the problem, of course, involves more than money. Developers have no interest in building for the poor. Neither do local and national leaders.

Squatters are neglected and disrespected by governments, politicians, the press, and even much of the public. And, like Elocy, squatters often neglect and disrespect themselves, as well. They judge themselves as useless, as parasites, as drains on society, as people who don't deserve a seat at the policy-makers table or a piece of the political pie.

In their early days, most squatter communities remained furtive, existing under the political radar. Indeed, this was their principal survival strategy: to build their homes on undesirable turf, places that allowed them to disappear from public view. But, as globalization has pushed the world's cities into frenzied competition for international tourist and development dollars while at the same time forcing more people to migrate to the cities, concealment is no longer an option. Now, if they are to secure their homes, squatters must assert themselves in a world that wants to deny their legitimacy and, in the most extreme cases, deny them the right to exist altogether.

To challenge this, squatters will have to mobilize and organize. They will have to learn how to engage the political system, how to strategize, how to take risks, and how to assess which risks to take. They will have to tap the strength they already have but don't yet see in themselves.

—RN, Brooklyn, NY. December 2005

Prologue: Crossing the Tin Roof Boundary Line

Let the wall crumble on which another wall is not growing.

—Cesar Vallejo

Tema said it with a sigh. He spoke softly with great fatigue, as if he was confiding something inexpressible, something sad, something he feared an outsider might never understand. I made him repeat the words: "Ai, Robert, o terceiro mundo é um jogo de video." "The third world is a video game."

It was around midnight. We were sitting in Beer Pizza, a restaurant halfway up the Estrada da Gávea, the main drag of the illegal neighborhood called Rocinha, the largest squatter community in Rio de Janeiro. The neighborhood was boogying. There was a convivial crowd at the outdoor tables of the pizzeria, and a guitarist had set up at one side of the courtyard. He sang bossa nova, Motown, and rock 'n' roll standards. Inches away, just beyond the curb, cars and buses and motorcycles jammed the roadway. A continual flow of people moved along the street. Scores of stores were still open, despite the hour. Just down the hill, six men were drinking *cachaça* and singing *pagode* at a small bar. One strummed a banjo while the others hammered the soft syncopated beat on their chairs as they sang. A few hundred paces farther up the slope, a dozen kids were playing soccer on a floodlit field, oblivious of everything around them except the black and white ball.

And then there were the homes. Little more than a decade ago, people here lived in waterlogged wooden barracks. When they wanted electricity, they stole it, looping long strands of wire through the trees and pilfering weak current from faraway poles. They hauled water up the hill in buckets and wheelbarrows and sometimes on the back of a burro.

But that is all in the past. Today there are thirty thousand homes in Rocinha spread across the sharp incline of Two Brothers Mountain. Most are two, three, or four stories tall, made from reinforced concrete and brick. Many boast shiny tile facades or fantastic Moorish balustrades or spacious balconies. which look out over the endless waves crashing on the beach at São Conrado, far down the hill. Electricity and water have come to this illegal city, and with them a degree of consumerism. Most families have a refrigerator, a color television (Jerry Rubin would approve), and a stereo. Rocinha today

is a squatter village 150,000 people strong—the largest in Rio de Janeiro. It occupies its hilltop redoubt between the wealthy neighborhoods of Gávea and São Conrado with the confidence of a modern, self-built Renaissance hill town.

One-fifth of Rio lives like this. A million people. They don't own the land, but they hold it. And no one contests their possession. Their communities are called favelas.

I reveled in the contrasts. Smokey Robinson and samba. A sidewalk cafe in the squatter neighborhood. Illegal houses with the best views in town. Permanent buildings in an impermanent community.

Yes it is a video game: the Marvelous City presented as a city of marvels, with a play of images and sounds as bright and diverting as in any Play Station or X-Box program. But for Tema, for the hordes on the hill, it was life, not display. They built their illegal homes simply because they couldn't afford anyplace else to live. And from that humble origin, against all odds, they produced something complex and sometimes harsh and unruly. They produced a new city.

The hut was made of corrugated metal set on a concrete pad. It was a 10-by-10 cell. Armstrong O'Brian, Jr., shared it with three other men.

Armstrong and his friends had no water (they bought it from a nearby tap owner), no toilet (the families in his compound shared a single pit latrine), and no sewers or sanitation. They did have electricity, but it was illegal service tapped from someone else's wires and could power only one feeble bulb.

This was Southland, a small shanty community on the western side of Nairobi, Kenya. But it could have been anywhere in the city, because more than half the city of Nairobi lives like this—1.5 million people stuffed into mud or metal huts, with no services, no toilets, no rights.

Armstrong explained the brutal reality of their situation. They paid 1,500 shillings in rent—about $20 a month, a relatively high price for a Kenyan shantytown—and they could not afford to be late with the money. "In case you owe one month, the landlord will come with his henchmen and bundle you out. He will confiscate your things."

"Not one month, one day." His roommate Hilary Kibagendi Onsomu who was cooking *ugali* the spongy white cornmeal concoction that is the staple food in the country, cut into the conversation.

"We kneel before the landlord and his agent all the time," Armstrong said.

They called their landlord a *wabenzi*—meaning that he's a person who has enough money to drive a Mercedes-Benz. He lives in a wealthy area, a community called Karen, in honor of Danish Baroness Karen Blixen who once owned a coffee farm there. Blixen left Kenya better than 70 years ago, when it was still a British holding, but her memory lives on in a book—*Out of Africa,* written under the pen name Isak Dinesen—and in that shaded grove of colonial entitlement on the edge of the Ngong Hills where her manor was located.

Hilary served the *uqali* with a fry of meat and tomatoes. The sun slammed down on the thin steel roof, and we perspired as we ate. After we finished, Armstrong straightened his tie and put on a wool sports jacket. We headed into the glare.

Outside, a mound of garbage formed the border between Southland and the adjacent legal neighborhood of Langata. It was perhaps 8 feet tall, 40 feet long, and 10 feet wide, set in a wider watery ooze. As we passed, two boys were climbing the Mt. Kenya of trash. They couldn't have been more than 5 or 6 years old. They were barefoot, and with each step their toes sank into the muck, sending hundreds of flies scattering from the rancid pile. I thought they might be playing King of the Hill. But I was wrong. Once atop the pile, one of the boys lowered his shorts, squatted, and defecated. The flies buzzed hungrily around his legs.

When 20 families—one hundred people or so—share a single latrine, a boy pooping on a garbage pile is perhaps no big thing. But it stood in jarring contrast to something Armstrong had said as we were eating—that he treasured the quality of life in his neighborhood. For Armstrong, Southland wasn't constrained by its material conditions. Instead, the human spirit radiated out from the metal avails and garbage heaps to offer something no legal neighborhood could: freedom.

"This place is very addictive," he had said. "It's a simple life, but nobody is restricting you, nobody is controlling what you do. Once you have stayed here, you cannot go back." He meant back beyond that mountain of trash, back in the legal city of legal buildings with legal leases and legal rights. "Once you have stayed here, you can stay for the rest of your life."

Sartaj Jaipuri was evicted in 1962, pushed out of Bombay's seaside Worli neighborhood because the government had determined that it would be the city's next commercial center.

He vowed he would never be booted from his home again. So he relocated to a place he thought would be safe. It was a dozen miles further out of town, far from the sea, far from the center of the city He moved his family to a steep unused plot near the tracks of the Western Railway in a scantily developed area called Malad.

It was rough living, but it was home. Sartuj and his fellow land invaders built their houses from bamboo topped with grass mats. The jungle was their toilet. They carried water from the public taps near the train station, a kilometer or so away. They christened their new community with an admirably straightforward name: Squatter Colony.

Squatter Colony developed with caution. The residents maintained a low profile for nine years before they took the risk of laying permanent foundations for their homes. Those who had money ripped out their original wood and mud platforms and laid down a brick base for their bamboo huts. Then they were quiet again for another decade, before they finally pooled their savings and paid a contractor to run water pipes and open communal taps. A few years later, they made another investment, again hiring the contractor to run the pipes directly into each home. In 1989, 27 years after they seized the land, they finally built something more permanent than their bamboo homes. They tore down the structures and the foundations and built anew with steel and concrete. They waited seven more years for the final piece of the puzzle—electricity.

Today there are perhaps a thousand families in Squatter Colony. Their homes are permanent and some are quite spacious. Most have water and toilets built inside. Sartaj's townhouse is on the upper end of a narrow lane that is paved with tiles and cement. His home, though on a tiny plot, is built to maximize space. The ground floor does quadruple duty as kitchen, living room, bedroom, and bath. There's a steep staircase that leads to a mezzanine, used for storage or an extra bed and on to a top floor where his youngest son, Aasil, bunks. Another son, Aarif, lives a few blocks down the hill, in a spacious, airy second-floor studio apartment. He's also a squatter.

"These houses are all illegal," Sartaj said. "Even where you are sitting right now is illegal." A slight, soft-spoken man who, among other professions, is a poet and lyricist, he sat cross-legged on the floor of his son's room and swiped one hand through his twist of white hair. He seemed, suddenly, too fragile and fatigued to be a homesteader. He sensed my skepticism and confronted it head-on: "These houses are made by us, by money of our own, and not by the government," he declared.

Mumbai as the city has been called since 1996, is India's richest city. The city's metropolitan area accounts for 40 percent of the tax revenues of the entire nation. Yet approximately half the inhabitants—more than six million people—have created

their homes the same way Sartaj Jaipuri did. They built for themselves on land they don't own. Mumbai is a squatter city. Still, Malad has gentrified over the years and land has become valuable. After more than 40 years in the home he built with his own hands, Sartaj Jaipuri finds himself wondering whether the future could be like 1962 all over again.

Yahya Karakaya came to Sultanbeyli in 1969. He was 4 years old, and all he remembers is a sleepy community of two dozen families in a wooded valley on the Asian side of Istanbul. The villagers raised cows, sold the milk to passing city-dwellers, and harvested lumber from the vast forest around them.

Today, Sultanbeyli is an independent squatter metropolis—population 300,000—and Yahya Karakaya is its popularly elected Mayor. From an oversized desk in a cavernous office on the seventh floor of the massive squatter City Hall, he presides over an empire that includes everything you thought squatters could never achieve: a planning department, a department of public works, a sanitation department, even a municipal bus service.

In Sultanbeyli nobody owns, but everybody builds. Fatih Boulevard, the main drag, is 5 miles long and boasts a strip of four-, five-, and six-story buildings complete with stores, restaurants, banks, and real estate brokerages. This illegal city even has its own post office.

With this level of development, Sultanbeyli has taken the quaint notion of squatter construction to a new level, For years, Turkey's squatters built at night to take advantage of an ancient legal precept that said, essentially, that if they started construction at dusk and were moved in by sunrise without being discovered by the authorities, they gained legal standing and could not be evicted without a court fight. That's why squatter housing in Turkey is called *gecekondu* (the "c" in Turkish is like "j" in English, thus: geh-jay-kon-doo) meaning "it happened at night." Half the residents of Istanbul—perhaps six million people—dwell in gecekondu homes.

In Sultanbeyli, the squatters are no longer furtive. Gone are the nights of anxiety and sweat as families built undercover of darkness. Gone are the tiny homes, designed to be erected quickly and to be hidden in sunken lots in order to escape official notice. Squatters in Sultanbeyli boldly proclaim their existence. Construction goes on in the open. 24 hours a day, "We are not gecekondu," the Mayor said with a smile. "We are gunduzkondu"—happening during the day.

Four cities. Four countries. Four continents. Four cultures. One reality: squatters.

Estimates are that there are about a billion squatters in the world today—one of every six humans on the planet. And the density is on the rise. Every day, close to two hundred thousand people leave their ancestral homes in the rural regions and move to the cities. Almost a million and a half people a week, seventy million a year. Within 25 years, the number of squatters is expected to double. The best guess is that by 2030 there will be two billion squatters, one in four people on earth.

As you might expect, with numbers like these, squatters are a pretty diverse bunch. There are those we are used to in the developed world, who intrude into buildings that are abandoned by their owners. There are those who build cabins in remote areas, farming land they don't own. There are those whose invasions are organized by a political outfit, like the Movement of Landless Workers, which is challenging the rule of the land barons in rural Brazil.

But these people are not the mass of squatters. The overwhelming majority of the world's one billion squatters are simply people who came to the city, needed a place to live that they and their families could afford, and, not being able to find it on the private market, built it for themselves on land that wasn't theirs. For them, squatting is a family value.

These squatters mix more concrete than any developer. They lay more brick than any government. They have created a huge hidden economy—an unofficial system of squatter landlords and squatter tenants, squatter merchants and squatter consumers, squatter builders and squatter laborers, squatter brokers and squatter investors, squatter teachers and squatter school kids, squatter beggars and squatter millionaires. Squatters are the largest builders of housing in the world—and they are creating the cities of tomorrow.

Three hundred people a day make the trek to Istanbul, three hundred more to Mumbai and three hundred also to Nairobi. Nicodemus Mutemi was one of them. He came to Kenya's capital in 1996 from his family's home in the Mwingi district. The Mutemi family cultivates corn and millet on their small holding in the parched hills an hour's walk from the nearest village. The land is dry in Mwingi—locals call it semiarid—and the air is still and hot. Growing crops in the cracked earth is a struggle. The family supplements its subsistence agriculture with a small herd of goats and a group of chickens and roosters.

Nicodemus's father poured some home-made honey beer from his gourd into a well-used plastic container. The brew was slightly sour and amazingly refreshing in the heat. As the sun tilted toward the horizon, slipping behind the silhouette of a baobab tree, Nicodemus explained why he left his homeland and clan and moved to Nairobi.

The problem, he said, is economic: You can grow enough to eat, but you can't grow enough to live.

Nicodemus hefted a burlap bag half filled with corn. That bag, he told me, would fetch five shillings at a local wholesale market. But to buy the corn back, in the form of *unga*, the flour used to make *ugall*, would cost 4.5 shillings at the local store.

The farm economy doesn't work. A farm family can raise enough to eat, but the crops alone will not generate an income. So how will the family members buy clothes or water or school books? How will they pay for kerosene or paraffin so they can light a lamp at night? How will they get tea for breakfast? And what about greater expenses? How will they repair the ancient mud and thatch huts that have served for generations but are beginning to crumble? And, if someone in the family gets sick, how will they pay for a doctor when medicine is a cash business.

The Mutemi family struggled to give Nicodemus an education. He graduated from Form 4—the equivalent of gaining a high school diploma. He would have liked to go to college, but there was no more money. Thus it became his turn to provide for his family, to repay his parents' investment, to secure a future for his own children. So he came to the city.

To be fair, Nicodemus's story is nothing new. This massive migration from rural regions to the urban centers of the world has been going on for thousands of years. And always, once they got to the cities of their dreams, the migrants have become squatters.

In Ancient Rome, despite the astounding government investment in public works, waterways, and infrastructure, squatters took over the streets, occupied fountains, and erected crude lean-tos called *tuguria,* tucked up against the sides of buildings. They were brazen and often seemed to dare authorities to remove them, but there were so many of them that the government couldn't keep up. And it has been like this in almost every city. Some sections of London were squatter zones until the mid-1800s. Paris, too, had its squatters, and historians suggest that the Court of Miracles, immortalized by Victor Hugo in *The Hunchback of Notre Dame,* was originally a squatter colony. Even New York, the definition of the modern real estate city, was a squatter metropolis until the early years of the twentieth century. In fact, the word *squatter* is an American term, originating in New England around the time of the revolutionary war as a popular term for people who built their homes on land they didn't own. The first use of the word in writing came in 1788, by the man who would become the fourth president of the United States—James Madison.

At the same time Nicodemus was establishing himself in Nairobi, I was beginning my own journey.

It was 1996 and the United Nations Commission on Human Settlements—Habitat, for short—the world body that studies and works on housing issues, was holding a major conference in Istanbul. Habitat holds these meetings once every decade, giving bureaucrats and nonprofit organizations a chance to compare notes and promote enlightened policies.

Preparing for that meeting, the statistics fell into place. If seventy million people are coming to the cities every year, and neither governments nor private builders are prepared to handle the onslaught, then all the government bureaucrats and staffers from nonprofits who were gathered in the fancy hotels overlooking the Bosphorus were in the wrong place. They should have been in Sultanbeyli and other squatter neighborhoods, learning from the land invaders.

I began to wonder about the morality of a world that denies people jobs in their home areas and denies them homes in the areas where they have gone to get jobs. And I began to think about my responsibility. I have written scores of articles on real estate, housing, architecture, design, business, planning. Wasn't I guilty, too? Hadn't I focused too much on developers and tycoons and architects, people who, despite the soaring ambition and ego contained in their buildings, have produced relatively little? Why wasn't I writing about the world's squatters, who have journeyed so far and produced so much without any noticeable self-aggrandizement?

After all, if society won't build for this mass of people, don't they have a right to build for themselves? And if they do then isn't there merit in their mud huts? If they are creating their own homes and improving them over time, then isn't there something good—at least potentially—about a community without water and sanitation and sewers? And if that's true, then shouldn't the comfortable class stop complaining about conditions in the shantytowns and instead work with the squatters to improve their communities?

They have created tiny ridges in the earth, outlines that indicate what is yours, what is mine. The dividing lines are nothing—scarcely more than an inch high, but pounded hard so they cannot be easily erased. Each seam delineates a living space. This is where people cook, read, eat, wash, sleep. This is where they store their food and their clothes. For the past three years, Laxmi Chinnoo, her mother, and her three daughters, have lived in one of these imaginary homes, under a bridge that crosses the tracks of the Harbor Line Railway not far from the Chuna Bhatti station in Mumbai.

Aside from those lines in the dirt and a few rugs hung on ropes so her daughters have a private place to change clothes, she has not built anything. There are a dozen other families living here in the same circumstances.

Are these people squatters?

Or how about Gita Jiwa, a construction laborer who has lived with her three daughters in a makeshift bamboo and plastic tent on the median strip of Mumbai's Western Expressway for the past five years? Fifteen families live alongside her. Are they squatters?

Or how about Washington Ferreira who lives with his mother and younger sister in a two-room rental in Rocinha. They are tenants, not invaders. Are they squatters?

To me, they are all squatters. But their experiences reveal that there are many different types of squatters, with different needs, different incomes, different aspirations, different social standing, different stories.

I'm standing on a wasteland. Several hundred acres, vacant, home only to scrub and weeds and illegally dumped trash. In the fall, the wind whips across these desolate blocks and the air turns tart against your skin. In winter, the flat expanse becomes a tundra as ice crusts the top of the construction debris. In springtime, butterflies squat on the tufts of sand grasses, and the land seems alive with possibilities. On bright summer days, dragonflies sprint above the cracked pavement, seeming to be racing their own shadows. This is beachfront property, perhaps ten miles from the tip of Manhattan, a bit more than an hour away by subway—Sprayview Avenue on the Rockaway Peninsula in Queens. Fifty years ago, it was a bungalow community—a summer resort for the lower middle class. Then, in the 1960s, the government took it for urban renewal. It has been vacant ever since. The paved streets, the rusting hydrants, the sewers, the streetlights—all the services people could need—have been in suspended animation, waiting for someone, anyone, to see the possibility.

Every time I visit Sprayview Avenue, I think of the third world. I think of Rio and Nairobi and Mumbai and Istanbul. In each of those cities Sprayview Avenue would have life. People who needed it would have seized the land and built their rustic homes. They would not be anarchists or radicals or hotheads. They would not be people with a political ax to grind or an ideological agenda. They would, rather, be regular people. Working people. People with families. With young children. People who came to the city to find work. Mechanics and waitresses, laborers and sales clerks, teachers and taxi drivers. These city-builders would construct using the crudest materials—mud, sticks, scavenged cardboard, wood, plastic, and scrap metal. At the start, their Sprayview Avenue would be a severely unhygienic place. No water. No toilets. No sewers. No electrical connections,

Eventually, though, one resident would have seen the potential and opened a bar by the beach, selling beer out of buckets of chopped ice. Another enterprising squatter would have started a restaurant—perhaps a pizza joint. Various small-scale entrepreneurs would have fashioned home-made pushcarts and plied the nearby boardwalk, selling *churrascos* or *nynma chnma* or *bhel puri* or *kofte*. A few years on, one canny fellow would realize that he could rent apartments at a nice markup (but still far less than in the surrounding legal neighborhoods) if he built with a degree of quality and style. So he would gather his neighbors, and they would rip down and build again, but this time with higher standards and nicer finishes. And then the neighborhood—self-built and self-governed but owned by no one—would have tenants, too.

Of course, we outsiders would find ways to discredit this free soil republic. We would call it a slum. We would warn our children: these are criminals, dirty people, thieves, muggers, prostitutes, gang leaders, disreputables, abusers. We would ignore the hard work it takes to build a community and argue instead that these are people trying to get something lor nothing, sponging off the system, ripping us off because they don't pay taxes. We would decry the density, the lack of adequate sanitation, the cacophony of construction styles, the sad-sack structural engineering. Politicians and real estate investors would call for inspections. Wealthy neighbors would clamor for police action. Together, we would make Sprayview Avenue a world apart. And ultimately, we would wipe it out.

Why do we have this animus against squatters? Why do we insist that there is something deeply wrong with their communities?

Favela kijiji, johpadpatti, gecekondu: Brazil, Kenya, India and Turkey have specific, descriptive, evocative terms for their squatter communities—in their own languages. It's the same around the world. From the *aashiwa'i* areas of Cairo to the *barriadas* of lima, the *kampungs* of Kuala Lumpur, the *mudukku* of Colombo, and the *penghu*, or straw huts, of Shanghai in the 1930s, most languages have specific and even poetic names for their squatter communities. But in English, there's come to be one dominant term: slum.

Why slum? By the dictionary, a slum is simply an overcrowded city neighborhood with lousy housing. But the term is laden with emotional values: decay, dirt, and disease. Danger, despair, and degradation. Criminality, horror, abuse, and fear.

Slum is a loaded term, and its horizon of emotion and judgment comes from outside. To call a neighborhood a slum immediately creates distance. A slum is the apotheosis of everything that people who do not live in a slum fear. To call a neighborhood a slum establishes a set of values—a morality that people outside the slum share—and implies that inside those areas, people don't share the same principles.

Slum says nothing while saying everything. It blurs all distinction it is a totalizing word—and the whole, in this case, is the false. So though it is the generally accepted term for squatter communities in both Kenya and India, I will avoid the word as much as I can.

I decided to do my part, to investigate the squatter communities of the world. At that point, each city made a case for itself.

Rio de Janeiro demanded that it be a focus because squatters there have a long and noble history. Their communities have existed for better than a century, and they have created permanent high-quality neighborhoods with high-rise buildings made from poured concrete and brick. Some of the city's squatter communities are so well-established that squatter houses command prices similar to those in legal neighborhoods of the city. Also, Rio's squatter areas have an impressive, dark subtext. For decades, national, state, and local governments steadfastly refused to provide services to these communities. And with that neglect came criminality. So most of Rio's favelas are now controlled by highly organized and extremely well-armed drug gangs. These gangs are both criminal and communitarian. They offer squatters a trade-off. In a city where assaults and violence of all sorts can be common, there is no crime in the squatter communities—as long as people look the other way when the dealers are doing their business. This, I thought, was an interesting story.

Nairobi claimed its place because two-thirds of its residents live in shantytowns and, in the 40 years since Kenya won independence from Britain the city's shantytown communities have remained unrelentingly primitive. What's more, Nairobi is the world headquarters of the UN's Habitat group, and I wondered why the agency had been unsuccessful in working to improve conditions for the 1.5 million people who live in the city's shantytowns—without water, electricity, sewers, or sanitation—just a few miles from its comfortable headquarters.

Mumbai insinuated itself because of its massive squatter presence. So many squatters live in the city that they have distinct class differences. Pavement dwellers—people who live in shacks built right on the sidewalks—are at the lowest end of the economic spectrum. People like Sartaj Jaipuri are at the higher end. Mumbai also boasts the largest squatter community in Asia, a neighborhood called Dharavi, which is now being eyed by developers because of its central location. In addition Mumbai is where Jockin Arputham lives. A generation ago Jockin founded a small community organization of squatters. Today, that group has become a multinational nonprofit organization active in a dozen countries. No story of squatters can be complete without spending some time with Jockin.

Finally, Istanbul leaped to mind. I knew the city had been the location of the United Nations meeting on housing in 1996. But, I would come to learn. Turkey has two notable laws that give squatters legal and political rights, and thus the chance to build permanent communities. If Turkey's legal system were in place in all the countries I visited, squatters would be in much better shape around the world.

They all laughed. Six men laughing because I didn't understand their concept of land ownership.

We were in a teahouse in a dusty patch of Istanbul called Paşaköy, far out on the Asian side of the city. Here, the streets were dusty cuts hacked into the scrubby hills. Each home, too, was dusty, caked, it seemed, with red earth. Even the giant blue plastic water barrels that stood in front of each house were coated with dust.

The tea, the men joked, was exotic—it had come from far away, Sadik Çarkir, the teahouse owner, had hauled the water from a spring several kilometers away. As we spoke, several women strode down the street with live buckets in a wheelbarrow. They were making the run to the source.

"Tapu var?" I asked. "Do you have title deeds?"

They all laughed. Or, more accurately some laughed, some muttered uncomfortably, and some made a typical Turkish gesture. They jerked their heads back in a sort of half-nod and clicked their tongues. It was the kind of noise someone might make while calling a cat or a bird, but at a slightly lower pitch. This indicates, "Are you kidding?" or "Now that's a stupid question," or, more devastatingly, "What planet are you from, bub?"

I blundered on.

"So who owns the land?"

More laughter. More clicking.

"We do." said Hasan Çelik, choking back tears.

"But you don't have title deeds?"

This time they roared. And somebody—I forget who—whispered something to my translator: "Why is this guy so obsessed with title deeds? Does he want to buy my house?"

You can't talk about squatters without talking about property. But talking about property involves different issues depending on where you are in the world.

In the developed world—particularly in the United States—many people still view property in the same absolutist terms that William Blackstone, the famed legal commentator sketched out in the eighteenth century: Property, he wrote, is "that sole and despotic dominion which one man claims and exercises over the external things of the world in total exclusion of the right of any other individual in the universe." What a revealing statement: property and despotism standing shoulder to shoulder. It's a distressing thought. Still, the United States maintains a hard-core devotion to property rights and free markets, which many economists contend, are the roots of all our liberties.

Alexis de Tocqueville recognized this feeling during his mid nineteenth century trip around the new nation in North America. "In no other country in the world is the love of property keener or more alert than in the United States," he wrote, "and nowhere else does the majority display less inclination towards doctrines that threaten the way property is owned."

Peruvian economist Hernando de Soto has adopted this hard core attitude and advanced a hypercapitalist argument in favor of squatters. De Soto suggests that the countries of the developing world should legalize their squatters just as the United States legalized the settlers throughout the western states under the Preemption Act of 1841 and the Homestead Act of 1862. De Soto argues that giving squatters individual title deeds will liberate what he terms the "dead capital" inherent in their homes, and will automatically give them a place in the market economy.

It sounds so simple: send some law school–trained Johnny Appleseeds to trek through the cities of the developing world, handing out title deeds. Then step back and watch the communities blossom.

I wish it would work.

No doubt, some squatters would be able to access more money if they had title deeds. But the folks I met in Brazil, Kenya, India, and Turkey didn't go through the tremendous struggles of building and improving their homes to liberate their dead capital. They go through incredible privation and deprivation for one simple reason: because they needed a secure, stable, decent, and inexpensive home—one they could possibly expand in the future as their families grow and their needs change. And title deeds—so natural to those of us who live in the developed world—can actually jeopardize this sense of security by bringing in speculators, planners, tax men and lots of red tape and regulations.

This is in part why they laughed at me all over the world when I spoke of private property. They laughed in Brazil, when I asked who owned the land in Rocinha. They laughed in Kenya, when I asked who owned the land under the mud and steel huts of the sprawling shanty communities. And they laughed in India, too, when I asked who owned the marshland that today is Dharavi. They didn't laugh because they would turn down a title deed if it was offered. They laughed because private ownership is not their most crucial concern.

When squatters feel secure in their homes, they build, invest, and prosper—and they don't need a title deed to do so. Squatters in Brazil and Turkey have erected permanent buildings without title deeds. Squatters in India have created whole neighborhoods while knowing that the land is not theirs. They have accepted the unofficial lines that divide one person's home from another's. They buy sell and rent their buildings. They negotiate with each other over their future plans for their homes.

The medieval Jewish sage Rashi proclaimed that being for what it means to be human being—to act, to live, to do things, even the most mundane things, in the world is essentially having a standpoint, a position, a base of operations. A massive number of people around the world have been denied that right. So they have seized land and built for themselves. With makeshift materials, they are building a future in a society that has always viewed them as people without a future. In this very concrete way, they are asserting their own being.

We can learn from their example. The world's squatters offer a different way of looking at land.

Rather than treating it as an economic value, squatters live according to a more ancient notion: the idea that every person has a natural right, simply by virtue of being born, to have a home, a place, a location in the world. Their way of dealing with land offers the possibility of a more equitable city and a more just world.

The Cities of Tomorrow

It is recorded that at first their dwellings were humble, mere huts and shacks, built of wood gathered at random, the walls plastered with mud. The roofs came to a point and were thatched with straw. But now all houses have a handsome appearance and are built three stories high.

—Thomas More

There is no mud hut utopia. Even in the mythical city of Amaurot, capital of the island of Utopia, people had to build, struggle, work, and fight to achieve the republic of their dreams. It's the same in the real world. As I think about the time I spent in the four cities I have profiled in this book, I think of the hard work involved in squatting: The discipline required to improve your house one wall at a time, and sometimes simply one brick at a time. The love people have for their communities. The pride in creation. The hunger people have for their efforts to be taken seriously. The desire for government to wrestle with their issues. And the modesty squatters display about their achievements.

While I was in Mumbai, Haaris Shaikh, a writer for a Marathi-language news weekly, asked me an excellent question. We were talking about my experiences in the squatter communities and my feeling that squatters were maligned through bad press. He interrupted to ask: "But aren't squatters the enemy of civil society?"

For once, I was prepared.

Think about it, I said: Squatters make up half the population of Mumbai. If they organized and pooled their votes, they could control the communities. If just 1 in 10 squatters organized to demand city services and, when they weren't provided, decided to march on the central business district, the crowd would be 600,000 people. They could paralyze downtown. They could outnumber the police. They could take over the city for a time. They could run civil society, or at least win whatever demand they were articulating at the time.

But they don't do this.

No, squatters aren't the enemy of civil society. They are the most law-abiding people around. As Valeria Cristina told me in Rocinha, "People may be poorer here, but they pay their bills. In Flamengo, which has rich people, many didn't pay their bills."

If the rich and wellborn were treated as badly by governments as the squatters have been, there would have been a rebellion long ago. The miracle is that the world's squatters value civil society and want to find a way of working within the system. They are law-abiding outlaws, patriotic criminals.

I'm standing on a wasteland. Forlorn, empty, unhappy, the gloomy pilings of the boardwalk in the distance, the lonely roar of the ocean on the unkempt beach beyond. This is, still, Sprayview Avenue in New York. Slowly, now, developers are picking off the parcels here, taking them from the government and building market-rate housing.

How much more quickly things would have gone, and how many worthy people would have good homes, if we had learned from the squatters of the developing world and allowed Sprayview Avenue to be built according to their model.

The world's squatters give some reality to Henri Lefebvre's loose concept of "the right to the city." They are excluded, so they take. But they are not seizing an abstract right, they are taking an actual place: a place to lay their heads. This act—to challenge society's denial of place by taking one of your own—is an assertion of being in a world that routinely denies people the dignity and the validity inherent in a home. As Patrick Chamoiseau put it in *Texaco,* his richly imagined fictional squatter history of Martinique, "In City, to be is first and foremost to possess a roof."

For a time, thousands of squatters in the cities of the developed world possessed that roof. Their history—one of mistreatment and, ultimately, eviction from the land, which, in some cases, had been theirs for decades—is not simply a tale of woe. Understanding this squatter heritage means accepting that squatters exist and that their constructions are a form of urban development. Squatters have been extremely effective at clearing land and building on it. They've never had the might to defy the moneyed interests for long.

But their brief successes—even in a world that espouses ever more strict adherence to property rights—show that there's another way to look at land; one that values possession more than purchase, and that recognizes need as well as greed. For those moments when squatters succeeded, there was freedom of domicile in our cities.

In the middle of my stay in Rio, I met Sonia Rabello de Castro. She's a former attorney general in the Rio de Janeiro government who now teaches law at the state university there. I thought that, as a person who has spent her life applying and upholding the law, she'd be against squatters. But she had a different viewpoint. She said that the favelas are here to stay, that they have become permanent, and there's no sense in government opposing them. She'd like to see squatters empowered to run their communities. But, unlike many of the other progressive lawyers around town, she doesn't think squatters need laws or lawyers to achieve great things. "They already have enough laws to exercise their rights," she told me. "The solution lies within, not outside. They do not need title deeds. They simply need a

few simple rules to create their own self-governing bodies."

She wasn't talking about politics, and least not politics in the electoral sense. She wasn't talking about bringing court cases, although that might be something squatters could decide to do. She was talking about old-fashioned grass roots organizing. People getting together, deciding what they want, and then figuring out how to get it. It's messy. It's time-consuming. It's frustrating. People will have to make mistakes and learn from them.

But it just might work.

Here's an example, from halfway across the globe.

Leonard Njeru Munyi and Samuel Njeroge grew up in Korogocho, a shantytown on the west side of Nairobi. Like many Korogocho kids, they were dump pickers when they were young. Nairobi's garbage dump is right next to Korogocho: a valley with a pall of smoke hanging over it because of the perpetual rubbish fires. Njeru and Njeroge culled the refuse every day, looking for items they could sell for a few shillings. It was simple survival.

Korogocho is tough turf—tougher, I think, than Kibera. Many of the people here were kicked out of downtown to make way for a real estate development, and they are still angry.

Njeru and Njeroge are still working together. Now they are social workers. When I met them they were employed by the government to work with street kids. And their lives mirror each other's. They both are married. They both have, as they call it, "micro families," meaning that each has just one child. They share a post office box. They share a mobile phone. They share, it seems, life.

We walked through the alleys of Korogocho and through the neighborhood called Kisumu Ndogo ("little Kisumu"), where vendors were frying fish on the street and selling chunks to the crowds coming home from work. Night was falling and we stopped in the gathering dusk to interview an *mzee* (a respected older man) who has lived in Korogocho for more than 40 years. As we spoke, a crowd gathered to listen. Eventually, the *mzee* tired of the discussion and wandered off. The others took over. We had what Njeru jokingly called a perfect focus group. I stopped asking questions and listened. The debate was between landlords and tenants. It was about people's rights. It was about how privileged people take advantage of others to grab land. Some of the younger folk were uncompromising: they wanted land reform now. Others said that you can't just take a man's land away from him even if he doesn't officially own it without giving him some consideration.

Someone—in the darkness I could no longer tell exactly who was speaking—told a story. Working together with a local school, the people in their part of Korogocho (a neighborhood called Grogan) built a toilet for the community. But after some payoffs to the local politicos a man took control of the toilet. He turned each section of the toilet into a single room, and has rented them out to families. Thus the community lost a toilet to one man's greed.

There were no answers during that night of political education. There was shouting, laughter, hard words, anger, duplicity, sincerity, frustration, communication. But no answers.

But since when are there ever answers in a day? The way forward is not a straight line, and the questions raised by the squatters are not easy ones. How do you organize a community where property does not exist? Whose interests are most important? What kind of homes do people want? Where should any money that might be available be put: into buildings, into infrastructure, or into education? How much can people really afford to pay in rent? If they are renters, who should own the buildings? These are tough issues and require tough debate. The people of Korogocho clearly have the appetite, energy, and intelligence to step into the conversation. The question is why the UN, the Kenyan government, and anyone else who cares about these communities doesn't work with them to start it.

This is what Korogocho needs, what Kibera needs, and Rocinha and Sultanbcyli and Dharavi and Squatter Colony. More focus groups, more debate, more discussion, more conversation.

The squatters are ready.

Are we?

doors-leading-to, never doors-against; doors to freedom: air light sure reason.

—Joao Cabral de Melo Neto

The Coming Plague

By Laurie Garrett

Introduction

Laurie Garrett's *The Coming Plague: Newly Emerging Diseases in a World Out of Balance* is a forgotten gem. Almost 20 years old, it still contains some of the clearest explanations of the connections between our actions and the ecological reactions that adversely affect us. As you might guess from the book's title, Garrett's thesis is that the loss of "balance," which could be defined as respectful treatment of the planet, will lead to one or more plagues, with potentially catastrophic consequences for humanity.

One such plague is already upon us: Sometime in the 20th century, AIDS emerged from the African forests as a result of deforestation. Forest roads led to increased contact with monkeys, and AIDS "jumped species" from a monkey it had long infected, to humans. Lacking any previous exposure, we have evolved no defenses against AIDS; as a result, it is almost 100% fatal if untreated. About 30 years since the first cases were diagnosed in the 1980s, AIDS is a global pandemic that has infected and killed tens of millions.

AIDS, however, is an extremely slow-spreading disease that takes years, if not decades, to kill its victims. Ebola and Marburg, as Garrett describes in other sections of this book, are other deforestation-related, simian-borne diseases that can kill within days. These hemorrhagic fevers have occasionally infected forest dwellers since time immemorial, killing so rapidly that neighboring villages would be spared and global pandemic averted because victims would die before they could walk out for help. In a world of logging trucks and jet airplanes, it is possible an airborne hemorrhagic fever could kill a large percentage of humanity within weeks. This danger was briefly popularized in Richard Preston's nonfiction book *The Hot Zone* and the action movie *Outbreak*, starring Dustin Hoffman; but as happens with all dangers that journalists and filmmakers sensationalize, the fear of epidemic has since faded into obscurity.

The danger, however, has not similarly faded. Smallpox was another hemorrhagic fever that jumped species a mere couple of thousand years ago. Almost 100% fatal at first, with repeated exposures over the centuries, it became a mere disfiguring ailment in Europe. Europeans taken with "the pox" could expect to break out in sores all over their body, suffer horribly, and live out the rest of their lives with horrendous scars on the face and body. A terrible fate to be sure, but nothing in comparison to its effect on Native Americans, who had never been exposed to European diseases before the Spanish Conquest. By some estimates, more than 90% of Native Americans were wiped out by such diseases, not only by smallpox but by minor ailments like measles. A newly emergent, airborne hemorrhagic fever could kill off a majority of humanity.

Bubonic plague, or "the Black Death," is transferred from rats to humans via flea bites. It reduced Europe's population by a third in the mid 14th century. Yes, you read that correctly: One third of all Europe died of the plague in a few years! Imagine a third of everyone you know dying—friends, family, loved ones. In modern history, we have not known this kind of loss. In the developed countries post–World War II, my generation did not even know the minor epidemics so familiar to my parents' generation: Polio, whooping cough, and scarlet fever seemed to be things of the past. Nationwide and even global public health campaigns were a major reason for this. Now we find the first two of these dread diseases are returning, along with terrifying strains of tuberculosis resistant to all antibiotics. Indeed, Garrett wrote another book called *Betrayal of Trust: The Collapse of Global Public Health* to plead for renewed vigilance. However, another reason for the enormous reduction in disease in modern times is that we live inside multiple bubbles—bubbles of cheap food and fuel, but also of effective antibiotics, of abundant and clean water that facilitates personal hygiene, and of effective sewer systems that wash pathogens away (though Garrett aptly shows that there is no "away" in our crowded and globalized world). I call these bubbles because, as currently implemented, they are as unsustainable as the housing bubble of the 2000s: Antibiotic resistance is on the rise due to misuse by the global medical and agricultural complexes; clean water is disappearing as populations increase and groundwater is depleted; and sewage, which the developing world will never be able to afford to treat with expensive and inefficient developed-world technologies, is a great growth medium for pathogens.[1]

I have chosen one chapter from Garrett's tome. Aptly titled "Nature and Homo Sapiens," it explains how climate change, ozone depletion, biodiversity destruction, and sewage issues have increased human and animal disease by changing ecological relationships between humans, bacteria, and viruses. If you read carefully, you will also find frightening tie-ins between pathogens and many other aspects of the global sustainability crisis, including genetically engineered food.

This chapter (and most of Garrett's book) paints a rather frightening picture, but there is hope in the closing paragraphs of *The Coming Plague*:

> In the end, it seems that the American journalist I. F. Stone was right when he said, "Either we learn to live together or we die together."

---

1 Another disturbing development is the emergence of an activist, middle-class, anti-vaccination movement in the developed world. So far, with considerable research, no connection has been shown between vaccinations and increased rates of child autism—but it is painfully clear, with the return of entirely preventable diseases like whooping cough, that a modern epidemic could be set off by mass vaccination refusals. This possibility will not be reduced in the developed world by revelations of CIA-sponsored vaccination programs designed to gain genetic information on suspected terrorists.

While the human race battles itself, fighting over more crowded turf and scarcer resources, the advantage moves to the microbes' court. They are our predators and they will be victorious if we, Homo sapiens, do not learn how to live in a rational global village that affords the microbes few opportunities.

It's either that or we brace ourselves for the coming plague.

So many books and movies end with a similar plea for human unity that it sometimes brings out the cynic in me. What, are we all going to have a global moment where we hold hands and sing "Kumbaya"? Well, no! But there has in the past been sane global collaboration on public health. Smallpox was the first and is so far the only disease to have literally been wiped off the face of the Earth.[2] The last case to occur was one Ali Maow Maalin, a cook at a hospital in Somalia. This was October 27, 1977. Victories like this show what is possible.

Global public health can also be augmented by a holistic approach to disease that favors strengthening immune systems over chemical eradication of pathogens. Example: When you overcrowd food animals, they get sick. The solution to this is not to maintain the crowding and nuke them with antibiotics; the solution is to take the animals out of confinement and put them back on the farm where they can move around, breathe fresh air, and encounter pathogens in lower densities that stimulate, rather than overwhelm, their immune systems.

Garrett is one of those marvelous authors who ties it all together for us. It is terrifying to be shown the connections between disease, ecological destruction, natural disasters and failed states. However, it is also hope-inspiring, since to tackle one of these problems is to tackle them all.

It is hard to gain historical perspective on an event that is completely unlike any other we have seen before.
—Al Gore, *Earth in the Balance*, 1992

That humanity had grossly underestimated the microbes was no longer, as the world approached the twenty-first century, a matter of doubt. The microbes were winning. The debate centered not on whether *Homo sapiens* was increasingly challenged by microscopic competitors for domination of the planet; rather, arguments among scientists focused on the whys, hows, and whens of an acknowledged threat.

It was the virologists, and one exceptional bacteriologist, who started the debate in 1989, but they were quickly joined by scientists and physicians representing fields as diverse as entomology, pediatric infectious disease, marine mammal biology, atmospheric chemistry, and nucleic genetics. Separated by enormous linguistic and perceptual gulfs, the researchers sought a common language and lens

2. In the natural world, that is. Samples were kept for research and found their way into Soviet bio-warfare laboratories where they were "heated up" for maximum virulence, mass-produced, and made into missile warheads (Richard Preston, *The Demon in the Freezer*, Random House, 2002). With the collapse of the Soviet Union, material and expertise diffused away from the massive *Biopreparat* program which employed 50,000 at its height. Today other nations, including North Korea and China, are believed to have smallpox "research" facilities.

through which they could collectively analyze and interpret microbial events.

There had never really been a discipline of medical microbial ecology, though some exceptional scientists had, over the years, tried to frame disease and environmental issues in a manner that embraced the full range of events at the microscopic level. It was far less difficult to study ecology at the level of human interaction—the plainly visible.

There were certainly lessons to be drawn from the study of classical ecology and environmental science. Experts in those fields had, by the 1980s, declared that a crisis was afoot spanning virtually all tiers of earth's macroenvironment, from the naked mole rats that foraged beneath the earth to the planet's protective ozone layer. The extraordinary, rapid growth of the *Homo sapiens* population, coupled with its voracious appetite for planetary dominance and resource consumption, had put every measurable biological and chemical system on earth in a state of imbalance.

Extinctions, toxic chemicals, greater background levels of nuclear and ionizing radiation, ultraviolet-light penetration of the atmosphere, global warming, wholesale devastations of ecospheres—these were the changes of which ecologists spoke as the world approached the twenty-first century. With nearly 6 billion human beings already crowded onto a planet in 1994 that had been occupied by fewer than 1.5 billion a century earlier, something had to give. That "something" was Nature—all observable biological systems other than *Homo sapiens* and their domesticated fellow animals. So rapid and seemingly unchallenged was human population growth, the World Bank predicted that nearly three times more *Homo sapiens,* on the order of 11 to 14.5 billion, would be crowded onto the planet's surface by 2050. Some high-end United Nations estimates forecast that more than 9 billion human beings would be crammed together on earth as early as 2025.

The United Nations Population Fund spoke of an "optimistic" forecast in which the planet's *Homo sapiens* population "stabilized" at 9 billion by the middle of the twenty-first century. But it was hard to imagine what kind of stability—or, more likely, *instability*—the world would then face, particularly given that the bulk of that human population growth would be in the poorest nations on earth. By the 1990s it was already obvious that the countries that were experiencing the most radical population growths were also those confronting the most rapid environmental degradations and worst scales of human suffering.

Biologists were appalled. Like archivists frantic to salvage documents for the sake of history, ecologists scrambled madly through the planet's most obscure ecospheres to discover, name, and catalogue as much flora and fauna as possible—before it ceased to exist. All over the world humans, driven by needs that ranged from the search for wood with which to heal their stoves to the desire for exotic locales for golf courses, were encroaching into ecological niches that hadn't previously been significant parts of the *Homo sapiens* habitat. No place, by 1994, was too remote, exotic, or severe for intrepid adventurers, tourists, and developers.

At Harvard University, Dr. E. O. Wilson was one of the leaders of a worldwide effort to catalogue the world's species and protect as much of the planet's biodiversity as possible. He estimated that there were 1.4 million known species of terrestrial flora, fauna, and microorganisms on earth in 1992, and perhaps as many as 98.6 million yet to be identified. The vast majority of those unknown plants and creatures, he argued, were living in the world's rain forests. There the plentiful supply of rain, tropical sunlight, and nutrient-rich soil bred such striking diversity that Wilson found 43 different *species* of ants living on a single tree in the Amazon. Devoted biologists were literally risking their lives in a mad rush to identify the missing 10 to 98.6 million species, some

50 percent of which were thought to reside in the rain forests of Amazonia, Central Africa, and South Asia.

The pace of the loss was staggering—on the order, by UN estimates, of 1.75 million acres annually.

Whether supplying the highly profitable heroin and cocaine markets, which in the Andes was responsible for devastation of upward of 90 percent of the Colombian forest and only slightly less alarming percentages of the forests of Ecuador, Peru, and Bolivia, the fast-food beef consumption habits, or the coffee needs of the wealthy world, entrepreneurs of the developing nations were responding to all too present economic incentives when they destroyed their natural ecologies. Without competing economic incentives for protecting the ecospheres it seemed unrealistic to expect that local human beings would take meaningful steps to reverse or slow the pell-mell pace of deforestation.

Using Landsat satellite imagery that was enhanced to reflect geographic features that might be hidden in flat photographs, David Skole and Compton Tucker, of the University of New Hampshire and NASA's Goddard Space Flight Center, made computer estimates of destruction in the Amazon between 1978 and 1988. Six percent of the Amazon's upper canopy and 15 percent of its total forest mass had, they concluded, effectively been destroyed.

Though it was well known to biologists that tiny isolated pockets of dense vegetation surrounded by devastation couldn't support a diverse range of species, none of the prior ecosphere calculations had factored for such islets of forestry. When Skole and Tucker studied the Amazon, however, they realized that many areas looked like a checkerboard, with slashes and zigs and zags of devastation slicing the rain forest into ever-thinner islets bordered by constantly thickening swaths of desertification or development. Humanity didn't nibble into the forest from its edges; it built huge superhighways that plunged into the pristine center and side roads that bisected one subsection after another.

So, the two scientists concluded, about 15,000 square kilometers of Amazonia were being directly destroyed by human beings every year, but another 38,000 square kilometers were indirectly destroyed annually by the isolation and fragmentation process. That combined effect represented an annual forest loss of an area larger than the United Kingdom of Great Britain and Northern Ireland. It also implied that between 1978 and 1988 Amazonia effectively lost 15 percent of its productive forest.

When ecospheres were so severely stressed, certain species of flora and fauna that were best suited to adapt to the changed conditions would quickly dominate, often at the expense of less flexible competitors. The net result would be a marked decline in diversity. This could clearly be visualized when, for example, a tropical area was cut to make way for a golf course. Though the golf course was composed of flora and fauna, its range of diversity was strictly controlled by human beings. At the course's periphery Nature would constantly try to push its way back in, but the aggressive species were usually limited to the healthiest plants and animals. If humans ceased trying to control the golf course, those sturdy aggressor species would swiftly move in, but it would be years before the original scope of diversity would be restored—if ever.

Both deforestation and reforestation could, therefore, give rise to microbial emergence. If an ecology had been entirely devastated, and its eventual replacement species were of inadequate diversity to ensure a proper balance among the flora, fauna, and microbes, new disease phenomena might emerge.

Such was the case in 1975–76 in the Atlantic seaside town of Lyme, Connecticut. Like many New England coastal communities that dated back to the colonial era, Lyme was a quaint town of two-hundred-year-old buildings, birch trees, and homes interspersed with pockets of picturesque pastoral scenery.

During the mid-1970s fifty-one residents of the town came down with what looked like rheumatoid arthritis. The ailment, dubbed Lyme disease, would by 1990 have surfaced in all 50 states and parts of Western Europe. Though scattered reports of Lyme would emanate from states with ecologies as disparate as those of Alaska and Hawaii, more than 90 percent of all cases were reported out of coastal and rural areas between Long Island, New York, and Maine. New York would, by 1988, lead the world in Lyme diagnoses with 6.09 cases per 100,000 adults, and reported cases from the northeastern states would double every year between 1982 and 1990.

The typical Lyme disease patient suffered localized skin reddenings that were indicative of insect bites, followed days to months later by skin lesions, meningitis, progressive muscular and joint pain, and arthritic symptoms. Untreated, the ailment could be lifelong, leading to a range of neurological disorders, amnesia, behavioral changes, serious pain syndromes in the bones and muscles, even fatal heart disease or respiratory failure. Once physicians learned of Lyme, the disease was undoubtedly overdiagnosed in endemic areas of the Northeast, but there remained a clear upward trend in the United States in bona fide cases, and by 1992 Lyme was the most reported vector-borne disease in the country.

Most Lyme sufferers lived in wooded areas that were inhabited by common North American feral animals: deer, squirrels, chipmunks, and the like. Notably absent in these untroubled, quiet woods were the ancient predators, such as wolves, cougars, and coyotes. Keeping deer and small mammal populations in check had, in fact, become a major headache for affluent wooded communities all over North America.

In 1982, Dr. Allen Steere of Tufts University in Boston discovered that Lyme patients were infected with a previously little-studied spirochete bacterium, *Borrelia burgdorferi.* Subsequently he and other physicians showed that many of the dreadful symptoms of the disease were the result of the immune system's protracted battles with the microbe.

Scientists soon determined that the *Borrelia* bacteria were transmitted to people by a tick, *Ixodes dammini.* While the tick was happy to feed on *Homo sapiens,* its preferred lunch was deer blood, specifically that of the white-tailed deer then common to the North American woods. About 80 percent of all North American cases were linked to either residing in a deer habitat or hiking through such an area.

Harvard's Andy Spielman showed, however, that getting rid of the deer in a region didn't eliminate Lyme disease. While the incidence of the disease among human beings might decline, it didn't go away. Further, there was a seasonal periodicity to Lyme outbreaks that coincided with the life cycle of the *I. dammini* tick, but not necessarily with that of the deer.

Spielman and his lab staff figured out that the ubiquitous northeastern mouse *Peromyscus leucopus* was the natural reservoir for the *B. burgdorferi* bacterium that caused Lyme disease. The immature ticks lived on the mice and fed on the rodents' blood. The mice, which were harmlessly infected with the bacteria, passed their *B. burgdorferi* on to the ticks. As spring approached, the winter thaw each year witnessed surges in the populations of both the *P. leucopus* mice and their tick passengers. The two species, rodent and insect, shared the ecology of low scrub brush that grew along the sand-duned shores and woodlands of the American Northeast. The deer grazed through these areas, picking up *I. dammini* ticks, which, while feeding on deer blood, passed on the bacteria.

The deer carried the ticks with them as they made long foraging journeys through woodlands and into suburban yards. Because there were no predators around to keep the deer population in check, their sheer numbers were great enough to force the animals to scour boldly for food, often stepping right into suburban front yards and patios to nibble at carefully cultivated azaleas and lawns. That, in turn, guaranteed that three more species—*Homo sapiens,* felines, and canines—would come in contact with *I. dammini* ticks and the *B. burgdorferi* bacteria they carried.

Studies in New York showed that the territory inhabited by the *I. dammini* tick was expanding at a steady and rapid rate, as deer, pet dogs, humans, rodents, and even some birds carried the insects further and further from the initial outbreak sites. By 1991, Lyme, the disease, and *I. dammini,* its vector, had spread widely throughout wooded and scrub-brush ecospheres all over the Northeast. Their invasion, and the epidemic they spawned, was new.

To understand how, and why, Lyme disease had suddenly emerged in North America, Spielman and his colleagues tried to recapitulate the history of the expansion of *I. dammini*'s territory.

The work took Spielman's group back in time to the arrival of British colonists in North America. When the Pilgrims landed in Massachusetts they set to work with Puritanical fervor clearing local forests and building settlements, Spielman said. By the late eighteenth century Massachusetts was the center of North America's iron industry, and remaining forests of the region were denuded to supply fuel for iron smelting. By the nineteenth century most of the woodlands of the entire Northeast had been so thoroughly devastated that housing construction required importation of wood from what were then the western territories.

"The result was an ecology just as artificial as a concrete parking lot," Spielman said, speaking of the later return of flora and fauna to the denuded areas. The grand tall trees, oaks and larches, never returned, nor did the large carnivorous animals. What did replace the old forests was an ecology similar to what probably had comprised the edges of the woods in the sixteenth century: scrub brush, small birches and other nonshade trees, meadows, deer, chipmunks, voles, squirrels, and birds.

"It's an artificial landscape that we have created, largely by neglect, here in the East," Spielman said, adding that the new ecology was filled with insect and rodent vectors, "lurking out there in this system of change."

Into the denuded forests came aggressor flora and, unchallenged by predators, the deer, rodents, and *I. dammini* ticks. As their numbers soared, bringing the deer, in particular, back from the edge of extinction in the Northeast, a new disease paradigm emerged.

As the invasion of *I. dammini* ticks and deer into artificially reforested areas demonstrated, no matter how hard *Homo sapiens* struggled to pave the world, Nature never ceased trying to force its way back. No area could escape the steady global spread of plant, animal, and insect species. In the absence of natural predators or competitors, alien species introduced into artificial ecologies—including mega-cities—could quickly overwhelm all suitable niches. And with the immigrant species could—and had—come microbes that were new to the local environment.

The Lyme case demonstrated the fallacy of viewing flora and fauna per se as "natural." From the point of view of microbial opportunity, loss of original biodiversity couldn't be corrected merely by introducing a handful of aggressor species.

During the early 1980s ecologists Paul and Anne Ehrlich of Stanford University developed the "Rivet Hypothesis" of diversity. They thought of the ecosphere as a huge airplane held together by steel rivets, or species. As each species died out, the total mass of the "airplane" might remain the same, but rivets were lost, weakening the overall structure. Eventually, a critical number of rivets having been lost, the plane would come apart, crash, and perish.

The epochal "Rivet Hypothesis" was given credibility by several experiments conducted in laboratories around the world. Scientists grew plants in environmentally sealed greenhouses filled with devices that measured carbon dioxide, oxygen, and total biomass. And it turned out that the more diverse the species assortment in a greenhouse—even when total biomass, sunlight, and all other factors were equal—the greater the oxygen production and general vitality of the little ecosphere.

In a survey of nineteen tropical forest ecospheres, researchers from the Missouri Botanical Garden found striking evidence that the changing ratio of oxygen to carbon dioxide was already having dire effects: forest turnover rates were increasing dramatically. Whole sections of forest biota "rivets" were dying and regenerating with radically escalating haste. In several major forests—particularly in Central Africa and Amazonia—turnover rates over the 1970–94 period had increased 150 percent every five years. The result, wrote Al Gentry (who died in a plane crash over Ecuador while making these surveys), was a net decrease in biodiversity as the older, massive hardwood trees, and the multitude of flora and fauna that existed in the ecospheres they created, died off and were replaced by a limited range of aggressive smaller trees and tropical vines. These gas-dependent species had less dense wood, and could transform forests into carbon sinks which would emit chemicals that further exacerbated the CO_2 imbalance and ozone crisis. Gentry predicted an accelerated rate of species extinctions and a radical change in the density and diversity of the world's rain forests, all occurring at astonishing speed.

From an atmospheric scientist's point of view the most crucial issue was the decline in oxygen production from the earth's flora due both to its overall declining mass and to the lowered range of diversity among surviving vegetation. Coupled with increased production of carbon dioxide owing primarily to human fossil fuel consumption and forest burning, and the expected increase in oxygen-dependent *Homo sapiens,* a clear chemical crisis loomed.

The most immediate impact was chemical destruction of the earth's ozone layer. The invisible layer of gas composed primarily of uncoupled oxygen atoms, or ozone, had unique physical properties. The atoms responded to specific wavelengths of light, repelling those in the ultraviolet and infrared bands. Little light in those wavelengths emanated upward from the earth's surface, but the planet was bombarded with such radiation from the sun. If not for the ozone layer, the planet would be a humanly uninhabitable hothouse bathed in mutation-causing ultraviolet light.

Throughout the 1980s researchers, particularly at NASA's Goddard Space Flight Center, amassed evidence that the ozone layer was weakening, especially over the South Pole. Over Antarctica an actual seasonal hole had developed in the ozone layer, through which poured levels of ultraviolet light unprecedented in known human history.

By 1990 a fierce debate raged in scientific circles over the size and significance of that ozone hole. But something was undoubtedly happening to the global ecology. Glaciers were retreating in some parts of the world, skin cancer rates were up in Australia and southern Chile, surface temperatures of oceans in some areas had risen, and mean surface air temperatures were up. Some researchers found, in fossils and deep glacial ore samples, evidence of such periods of warming in the earth's past, indicating that such events could all be part of a historic cycle on the planet. Further, it was possible that the bulk of the warming was induced not by human pollution and rain forest destruction, but by natural catastrophic events such as the 1991 eruption of the Mount Pinatubo volcano in the Philippines.

There was strong evidence, however, that halogen ions, particularly chlorides and bromides, were making their way via human pollution into the atmosphere. These were the breakdown products of thousands of plastics, pesticides, fuels, detergents, and other modern materials. Once inside the ozone layer, the halogens acted as chemical scavengers, attaching themselves to free oxygen atoms to form heavy molecules that then fell out of the protective layer into lower tiers of the earth's atmosphere. In this way, ozone was actively depleted.

Most Western scientists insisted that the pollution- and deforestation-driven ozone depletion and global warming hypothesis was correct, though among believers there were significant differences of opinion about its current and forecast severities. The strongest evidence supported an estimate of a net global temperature increase of half a degree centigrade during the twentieth century, with five degrees centigrade marking the difference between, on the one hand, the Ice Age and, on the other, a severely deleterious greenhouse warming effect.

The first outcome of this warming was a higher surface water evaporation rate, which, in turn, led to greater levels of rainfall and monsoon in key areas of the planet. In places that normally had low levels of rainfall, such as the Sahara Desert, there would be even less precipitation. The net result would be greater extremes in water distribution, with severe droughts afflicting some parts of the planet, flooding and hurricanes hitting others. That, in turn, was expected to alter everything from the migrations of birds to the feeding patterns of blue whales; from habitat ranges of malarial mosquitoes to the amount of the planet's arable land suitable for profitable agricultural growth.

The lesson of macroecology was that no species, stream, air space, or bit of soil was insignificant; all life forms and chemical systems on earth were intertwined in complex, often invisible ways. The loss of any "rivet"—even a seemingly obscure one—might imperil the physical integrity of the entire "plane."

The "plane," in the Ehrlichs' scenario, was destined to crash. What hadn't been anticipated was that the plane would first get sick, heavily encumbered by emerging pathogenic microbes.

In 1987, Siberian fishermen and hunters working around Lake Baikal noticed large numbers of dead seals (*Phoca sibirica*) washing up along the shores of the huge Central Asian lake. By year's end, the seal death toll would top 20,000, or nearly 70 percent of the entire population. The world's deepest lake—a mile deep—was a unique 12,000-square-mile ecosphere inhabited by a number of species of flora and fauna found nowhere else in the world, including the dark gray freshwater seals. Because the Soviet government had long used the country's lakes as waste dumps, it was first assumed that the seals were victims of some toxic chemicals.

But with the spring thaw of 1988 came an apparent epidemic of miscarriages among female harbor seals (*P. vitulina*) in the North Sea along the coasts of Sweden and Denmark. Some 100 spontaneously aborted seal pups were recovered, a few of which survived long enough for scientists to study their symptoms: lethargy, breathing difficulties, nasal congestion. A quick-and-dirty analysis of the pups' blood revealed that the dying and dead seals had antibodies that reacted weakly in laboratory tests against canine distemper viruses.

With the arrival of summer 1988 came hundreds of dead adult seals in the North Sea area. They washed up upon shores separated by huge expanses of land and sea, from the western Baltic Sea area of Sweden and Denmark to the far west coast of Scotland. By August dead seals were even found scattered along the beaches of northern Ireland.

In laboratories in the Netherlands, Ireland, Russia, and the United States, scientists swiftly determined that the seals were dying from a morbillivirus—the same class that included human measles, cattle rinderpest virus, and canine distemper. The die-offs continued well into 1990, eventually claiming more than 17,000 North Sea harbor seals, or more than 60 percent of the entire population.

Scientists working in laboratories from Atlanta to Irkutsk swiftly determined that two different viruses were responsible for what seemed to be separate seal epidemics in Lake Baikal and the North Sea. The virus isolated from the massively infected lungs of Lake Baikal's seals was dubbed phocine distemper virus-2 (PDV-2), and it proved virtually identical to canine distemper virus.

The North Sea seals, however, were suffering from something never before seen. Their microbial assailant, named phocine distemper virus-1 (PDV-1), was distinct from any other known morbillivirus. It appeared to be something new, and the extraordinary death rates among harbor seals indicated that their immune systems had never previously encountered such a virus.

While the seal experts worked on that puzzle, veterinarians in Spain were examining dolphins that were beaching themselves along the Mediterranean coast of Catalonia, Spain. By July 1990 more than 400 Mediterranean dolphins had washed onto the shores of North Africa, Spain, and France, clearly suffering respiratory distress. Autopsies of the animals revealed startling brain damage and acute immunodeficiencies. Similar signs of immune deficiency had already been documented in the North Sea seals, and accounts in the popular Spanish press were soon calling the mysterious marine mammal ailment "dolphin AIDS."

But it wasn't AIDS—it was more like measles. Dolphins were also coming down with a deadly morbillivirus. By 1991 common dolphins (*Delphinus delphis*), striped dolphins (*Stenella coeruleoalba*), white-beaked dolphins (*Lagenorhynchus albirostris*), and porpoises (*Phocoeona phocoena*) all over the Mediterranean and Ionian seas were dying.

Dutch scientists determined that at least four newly discovered viruses were attacking Europe's and Central Asia's marine mammals: PDV-1, which was similar to human measles; PDV-2, which appeared to be identical to the virus that caused distemper in dogs; and PMV or porpoise morbillivirus, and DMV, or dolphin morbillivirus.

Russian scientists discovered a possible explanation for the Lake Baikal epidemic of PDV-2. It seemed that an epidemic of distemper swept through the Siberian sled dog population in 1986; local people threw their dead dogs into the lake. Curious seals that investigated the large corpses became infected. In the Siberian case, then, the virus was not new to the world, though it was new to the freshwater seal species. The extraordinary death toll was the result of a microbe jumping from an ancient host species into a new, immunologically vulnerable species.

While the PDV-2 puzzle appeared to have been solved, mystery continued to shroud the origins of PDV-1, DMV, and PMV. Where did the viruses come from? How did they spread so rapidly over such a broad geographic area? Were they old viruses with newfound mutant virulence? Or were the animals particularly vulnerable because of other factors?

Spanish researchers were convinced that PCB pollution of European seas played a key role. All the dolphins they autopsied had high levels of PCBs stored in their body fat; some showed signs of PCB-induced tumors. So one hypothesis was that seals and dolphins that inhabited particularly contaminated waters were already immune-deficient when the virus appeared, making them uniquely vulnerable. Such an explanation might resolve questions about why Canadian seals, though infected with PDV-1, apparently hadn't fallen ill, while their cousins living in polluted North Sea and Baltic waters, had. But still unanswered was the origin of the viruses.

By 1993 the dolphin and porpoise die-offs had slowed considerably, and many scientists felt that the epidemic might be over. But why?

A key difference between 1988–90 and 1993 was the severity of Europe's winter. In the Mediterranean, in particular, the earlier winters were remarkably mild, which meant that small fish populations in the region were unusually low. Spanish researchers examined some 500 dolphin corpses, and concluded that all the animals were unusually skinny and their livers were severely damaged. The scientists decided that PCBs, which are normally stored in human and animal body fat, had flooded the dolphins' livers as the starving animals burned up stored body fat. That, in turn, led to high blood levels of the toxic chemicals and mild immune deficiency. The virus subsequently exploited the dolphins' vulnerability.

By the close of 1990 at least 1,000 Mediterranean and Ionian dolphins had succumbed.

The morbillivirus mystery deepened still further when bottlenose dolphins, beluga whales, Atlantic harbor seals, and porpoises were washed ashore on beaches stretching from the Gulf of Mexico to Quebec's St. Lawrence Seaway. Those that were discovered alive often appeared dazed and distressed, as if suffering brain damage or high fevers.

Again, scientists looked for viral and pollution explanations, finding a confluence of factors at play. As had been the case with European sea mammals, the North American die-offs came during unusual weather. An El Niño weather pattern gave rise to extraordinary rainfall in the Midwest, which led to high levels of pesticide, pollutant, and human and livestock fecal waste runoff into the Mississippi and other major arteries. That waste made its way to the Gulf of Mexico, where some of the first bottlenose die-offs occurred. The polluted waters moved through the Gulf, up the Florida coastline to the Carolinas, where more marine mammal die-offs ensued. From there, the water mass wended its way along the coasts of New Jersey, New York's Long Island, Massachusetts, Maine, the St. Lawrence Seaway, and Nova Scotia, everywhere claiming a toll.

Debates about what factors in that water led to the die-offs raged well into 1993. PCBs and other chlorinated hydrocarbon toxic chemicals were in the river runoff water, but few scientists believed the chemicals were directly responsible; many accepted the notion that chemically induced immunodeficiency served to aggravate some other underlying cause of disease in the dolphins, seals, whales, and porpoises.

Off the Carolinas scientists discovered massive colonies of algae of the species *Ptychodiscus brevis* that secreted a powerful neurotoxin called brevetoxin. They hypothesized that the unusual weather, coupled with high levels of nitrogen-rich human and livestock fecal matter, had led to the formation of huge "red tides," or algal blooms, that contained such toxic algae.

It was a tempting explanation, not only for the situation in the Americas but also for the European epidemic. Whether the animals were killed directly by some species of algae or indirectly by morbilliviruses that were spread around the world by hiding inside such algae, it would no longer appear terribly mysterious that seals in the Ionian Sea, Mediterranean, Baltic, North Sea, Gulf of Mexico, and off the shores of Long Island should all experience lethal epidemics at roughly the same time and under similar climatic conditions.

For more than two decades biologist Rita Colwell of the University of Maryland had been amassing evidence that bacteria and viruses lurked inside algae, and by the late 1980s other scientists were not only acknowledging the tremendous body of evidence but also correlating her algal findings with disease outbreaks in marine life and humans. Colwell knew that hundreds—perhaps thousands—of species of predatory algae were capable of secreting toxins designed to kill or paralyze their larger marine prey, allowing groups of the microscopic beings time to consume their conquest at leisure.

Algae were the oldest living species on earth, thought to have developed out of the planet's primordial soup more than three billion years ago. As creatures they resembled protozoa, but their use of chlorophyll to convert the sun's energy into useful chemicals made them, technically, plants. Algae clustered in both fresh and salt water, sometimes forming visible discolorations and "tides" on the surface. Three broad categories of algae were designated on the basis of their colors, which, in turn, reflected the nature of their internal chemistry: blue-green, red, and brown.

Algae could, during times of environmental stress or food shortages, encyst themselves in a protective coating, go dormant, and drop into hiding for extended periods. Once activated, however, algae needed sunlight and plenty of nitrogen. Most species preferred warmer waters, and under ideal conditions could multiply rapidly, drifting about in massive colonies that, in the case of oceanic red or brown tides, could span surface distances larger than Greater Los Angeles.

Just as E. O. Wilson speculated about the tremendous numbers of terrestrial species of all sizes that had yet to be discovered in the earth's rain forests, Colwell was concerned about the mysteries of the planet's marine world.

"Of the some 5,000 species of viruses known to exist in the world, we've characterized less than four percent of them," Colwell said. "We've only characterized 2,000 bacterial species, most of them terrestrial. That's about 2,000 of an estimated 300,000 to one million thought to exist. Less than one percent of all ocean bacteria have been characterized."

Colwell had devoted years to the study of microorganisms living in Maryland's Chesapeake Bay, where she discovered that viruses were seasonal: they reached their nadir in population during the icy months of January, when there were about ten thousand viruses per milliliter of water, and increased steadily in numbers as the bay warmed. By October, after three hot summer months, there were as many as a billion per milliliter, and the viruses outnumbered algae and bacteria. Even more profound variation in viral populations was seen in the waters of Norway's fjords, and Norwegian researchers were

convinced that viruses passed genetic material on to algae to assist in their adaptation to change.

In Colwell's Chesapeake some of the swollen summer viral population was indigenous to the bay, having simply multiplied in number as the water warmed. But increasingly over her more than thirty years of studying the Chesapeake, Colwell saw viral intrusion occurring, as human and animal waste washed into the bay, carrying with it a variety of pathogens. The greatest density of intrusion was around dump and sewage sites, where Colwell found veritable stews of viruses, plasmids, transposons, and bacteria intermingling.

"The probability of genetic exchange is very great," Colwell said. Indeed, lab studies had shown that some ocean bacteria possessed antibiotic-resistance genes, presumably acquired under just such conditions. Those newly antibiotic-resistant bacteria were, in turn, ingested by various mollusks and then eaten by sea mammals and humans. The mollusks and crustaceans—from scallops to lobsters—readily ingested all manner of microorganisms found along the world's polluted coastlines, including a host of enteric human pathogens.

"We have very few places left on earth where we can get pathogen-free mollusks," Colwell said.

Hepatitis, Norwalk virus, polio, and a host of other microbes were turning up in shellfish caught in the world's coastal waters, particularly around waste dump sites. And strange microbes appeared that burned through the shells of mollusks, killed off salmon, and made lobsters lose their sense of direction.

By one calculation a single gram of typical human feces contained one billion viruses. And in a liter of raw human waste there were more than 100,000 infectious viruses—none of which were vulnerable to mere chlorine treatment. Chlorine might eliminate the bacteria—though increasing chlorine resistance in bacterial populations was rendering such chemical sanitation insufficient—but viral elimination required more extensive filtration and tertiary treatment.

Ocean pollution due to raw sewage, fertilizers, pesticides, and other chemical waste was increasing steadily, producing tremendous changes in coastal marine ecospheres. Though the World Bank and the United Nations had designated sewage and sanitation systems a top priority for development during the 1980s, it was estimated that at least two billion *Homo sapiens* had no access to a sanitary fecal waste disposal system, most of them residents of Africa and southern Asia. Their fecal waste, as well as that of their domestic animals, ended up in nearby rivers, streams, and seas.

Algal blooms, as a result, increased in frequency and size worldwide throughout the four post–World War II decades. The nutrient supply provided by steady flows of fecal matter, garbage, fertilizers, silt, and agricultural runoff gave the algae plenty of food. Many scientists thought that the thinning ozone layer warmed the sea surfaces to temperatures suitable for microbial growth and reproduction. Algal blooms grew so rapidly on the surface of lakes, ponds, and the open sea that they actually blocked all oxygen and sunlight for the creatures swimming below, literally suffocating fish, marine plants, and mollusks. And some scientists believed there was evidence that the additional load of ultraviolet light making its way through the ozone layer was driving a higher mutation rate in sea surface organisms, possibly allowing for more rapid rates of adaptive evolution. If such a mechanism were in effect, it would favor microorganisms, which, on a population basis, were well positioned to make use of helpful mutations and tolerate individual die-offs due to disastrous mutations. The reverse would be the case for more complex marine creatures, such as fish, whales, and dolphins.

"The oceans have become nothing but giant cesspools," declared oceanographer Patricia Tester, "and you know what happens when you heat up a cesspool."

Jan Post, a marine biologist at the World Bank, used a similar metaphor when announcing the release of the Bank's 1993 report on the condition of the seas: "The ocean today has become an over-exploited resource and mankind's ultimate cesspool, the last destination for all pollution."

Tester, who worked for the U.S. National Marine Fisheries Service in Beaufort, North Carolina, had been monitoring weather patterns and algal blooms. She was one of many oceanographers who noted that the die-offs of dolphins, seals, porpoises, and whales during 1987–92 coincided with massive algal-induced bleaching of coral reefs worldwide and enormous red tides. She felt that there was compelling evidence for not only increased frequency and size of algal blooms but also their territorial expansion into latitudes of the seas formerly considered too cold for such algal growth. Using satellites to track the algal blooms, scientists documented increases—in some cases doublings—in size and scope of algal blooms during the 1980s and early 1990s.

Meanwhile, the overall diversity of the marine ecosphere was declining at a dramatic rate. With more than 95 percent of all marine life adapted to coastal regions, their susceptibility was high: human interference in the form of coastal development, sewage, and fishing was claiming a huge toll. U.S. Fish and Wildlife Service biologist Kenneth Sherman calculated that biomass production off the shores of New England, for example, had declined by more than 50 percent between 1940 and 1990 due, primarily, to overfishing.

A feedback loop of oceanic imbalance was thus in place. As the populations of plankton/algae eaters—whales, for example—declined, only the viruses remained to keep blooms in check. But raising the sizes of viral populations in the world's saline soup held out other dangers to marine and, ultimately, human health.

Rita Colwell was convinced that the entire oceanic crisis was already directly imperiling human health by permitting the emergences of cholera epidemics. During the 1970s she showed that the tiny resilient cholera vibrio could live inside of algae, resting encysted in a dormant state for weeks, months, perhaps even years. Colwell, a gritty, energetic woman, fought hard for years to convince the world's public health establishment that the key to forecasting emergence of cholera lay in tracking algal blooms that drifted from the shores of Bangladesh and India, key endemic sites for the microbe.

"But the bloody stupid physicians have this idée fixe that cholera is only directly transmitted, from person to person," Colwell said. "They just couldn't wrap their minds around the concept of microbial ecology. They fight me tooth and nail at every turn."

It was the emergence of cholera in Peru in January 1991 that compelled the World Health Organization and the global medical community to take notice of Colwell's message.

The global Seventh Pandemic of cholera began in the Celebes Islands in 1961, with the new strain, dubbed *Vibrio cholerae* 01, biotype El Tor. By the late 1970s the El Tor microbe had made its way into all the developing coastal countries of southern Asia and eastern Africa, and it was impossible to control it. It would not be until the 1991 Peruvian outbreak, however, that WHO and health experts would publicly acknowledge what Colwell had been saying for years: namely, that the El Tor strain was particularly well equipped, genetically, for long-term survival inside algae.

Since the early 1980s Colwell had been collaborating with the International Centre for Diarrhoeal Disease Research in Dhaka, Bangladesh, eventually becoming its research chair. There, in the heart of cholera endemicity—perhaps the cradle of all cholera epidemics—Colwell and her colleagues discovered that the El Tor strain was capable of shrinking itself 300-fold when plunged suddenly into cold salt water. In that form it was the size of a large virus, very difficult to detect. But the presence of hibernating cholera vibrio in a water source, or inside algae, could be verified by simply taking a sample and, in the laboratory, changing the conditions: add nitrogen, raise the temperature, decrease the salinity, and bingo! instant cholera.

They further discovered that the vibrios could feed on the egg sacs of algae: up to a million vibrios were counted on the surface of a single egg sac. That explained why health authorities couldn't manage to eliminate El Tor once it had entered their communities. The organism simply hid in algal scum floating atop local ponds, streams, or bays, lurking until an opportune moment arrived for emergence from its dormant state.

When El Tor hit the coastal parts of Peru in early 1991 the country was caught completely unprepared for such an occurrence. In Peru's hot summer January—made hotter still by an El Niño event—a Chinese freighter arrived at Callao, Lima's port city. Bilge water drawn from Asian seas was discharged into the Callao harbor, releasing with it billions of algae that were infected with El Tor cholera.

The first human cases of the disease offered the microbes terrific opportunities for spread in Peru. A national summertime delicacy was ceviche, or mixed raw fish and shellfish in lime juice. The bilge-dumped vibrio had quickly infected Peru's shellfish, so uncooked ceviche represented an ideal vehicle of transmission.

The second ideal opportunity for transmission of the microbes was provided by Lima's largely unchlorinated water supply. Because of both cost constraints and U.S. Environmental Protection Agency documents that indicated there was a weak connection between ingesting chlorine and developing cancer, Peru had abandoned the long-standing disease control practice of using the chemical to disinfect public drinking water. Later, CDC studies would show that the majority of Peru's cholera microbes were transmitted straight into people's homes, dripping from their water faucets.

The first cholera cases hit Lima hospitals on January 23; days later cholera broke out some 200 miles to the north in the port town of Chimbote.

As the El Niño water spread out along the Pacific coast of the continent carrying with it bilged algae, cholera appeared in one Latin American port after another. Eleven months into the Western Hemisphere's pandemic, cholera had sickened at least 336,554 people, killing 3,538. Throughout those months the microbe's emergence was aided by obsolete or nonexistent public water purification systems, inadequate sewage, and airplane travel. Cases reported in the United States involved individuals who boarded flights from Latin America unaware that they were infected, and fell gravely ill either in flight or shortly after landing.

Colwell and her colleagues demonstrated that the vibrio in the algae and those recovered from ailing patients were genetically identical. Further, they showed that the El Tor substrain, Inaba, which was raging across Latin America, possessed genes for resistance to the antibiotics ampicillin, trimethoprim, and sulfamethoxazole. The same substrain was highly antibiotic-resistant in Thailand, where it was invulnerable to eight drugs.

In Latin America the epidemic raged on well into 1994, with, according to WHO officials, "no end in sight." More than $200 billion would be spent by Latin American governments by 1995, according to the Pan American Health Organization, for emergency repairs of water, sanitation, and sewage systems. Only about 2 percent of all cholera cases were actually reported to authorities, WHO said, and across the continent 900,000 cases were officially reported as of October 1993, involving more than 8,000 deaths. Officially reported numbers of cholera

cases were so grossly understated that, by 1994, the only accurate statement one could make was this: between January 1991 and January 1994 millions of Latin Americans fell ill with cholera, thousands died, and the epidemic continues.

Once chlorine was vigorously introduced into Peruvian water supplies, the 01 strain proved fairly resistant to the chemical.

Though it was obvious to scientists all over the Americas by 1992 that the El Tor epidemic had succeeded in becoming *endemic* cholera in much of Latin America largely because the microbe was carried in algae, the real challenge to rigid old analyses of the spread of the vibrio came in December 1992 when an entirely new strain of cholera emerged in Madras, India. Dubbed Bengal cholera, or *V. cholerae* 0139, the newly emergent microbe competed with El Tor for control of the Bay of Bengal ecology. By June 1993 Bengal cholera had claimed over 2,000 lives and caused severe illness in an estimated 200,000 people. It had spread across much of the coastal region of the Bay of Bengal, encompassing the Indian metropolises of Calcutta, Madras, Vellore, and Madurai, as well as most of southern Bangladesh.

This new Bengal cholera appeared to be spreading far faster than the Seventh Pandemic. It look three years for that cholera strain to spread from India to Thailand, but the Bengal cholera had already turned up in Thailand's capital, Bangkok, by mid-1993, and threatened to spread nationwide, according to researchers from Mahidol University in Bangkok.

In March the leading hospital in Dhaka was healing 600 Bengal cholera cases a day—three times their normal daily cholera rate. In rural parts of Bangladesh cholera victims were reportedly falling ill at rates up to ten times those seen with the previous year's classic cholera outbreak.

Prior to the Bengal cholera outbreak there were two types of cholera in the world: classic and El Tor.

Classic cholera, which was endemic in parts of India and Bangladesh, was extremely virulent and easily passed from one person to another via contact with microscopic amounts of feces. The El Tor type, in contrast, was less virulent but could survive in the open environment far longer. A hallmark of the El Tor strain was its ability to move in the open oceans, as a silent passenger inside algae.

The Bengal cholera appeared to represent a combination of characteristics found in both the El Tor and the classic vibrio. Researchers from the International Centre for Diarrhoeal Disease Research in Bangladesh reported that the new mutant "may be hardier than and probably has survival advantage over" the classic strain of the bacteria. They found thriving colonies of the Bengal organism in 12 percent of water samples they tested, and the bacterial toxin was in 100 percent of all waters examined in Bangladesh.

One genetic trait was clearly missing in the new Bengal strain: that which coded for antigens that were usually recognized by the human immune system. As a result, people did not seem to have antibodies to the new mutant, and even adults who had survived previous cholera outbreaks appeared to be susceptible to the Bengal strain.

Genetic analysis of the new mutant vibrio suggested a terrible scenario: that it was essentially the El Tor strain possessing the virulence genes of classic cholera. As such, it would represent an entirely new class of cholera microbe, the like of which had never been seen. The emergence of 0139 "hit epidemiologists and physicians like a two-by-four between the eyes, because there is no explanation for its emergence and spread but ecology," Colwell said in the fall of 1993.

In 1993 Colwell teamed up with two Cambridge, Massachusetts, physicians to try to pull together the Big Picture, an explanation of how global warming, loss of oceanic biodiversity, ultraviolet radiation increases, human waste and pollution, algal blooms, and other ecological events joined forces. Together, they theorized that the cholera microbe defecated by a man in Dhaka, for example, got into algae in the Bay of Bengal, lay dormant for months on end, made its way via warm water blooms or ship bilge across thousands of miles of ocean, and killed a person who ate ceviche at a food stand in Lima. Drs. Paul Epstein and Timothy Ford, both members of a group of physicians and scientists at the Harvard School of Public Health calling themselves the Harvard Working Group on New and Resurgent Diseases, were convinced that essential to protecting their Boston patients in the twenty-first century was a better understanding of what was transpiring in the oceans. They saw a complex interplay at work, involving global climate changes, pollution, and the microorganisms.

In Epstein's view, algal blooms were giant floating gene pools in which antibiotic-resistance factors, virulence genes, and plasmids moved about between viruses, bacteria, and algae. He thought that ultraviolet radiation might be hastening the mutational pace. And terrestrial microbes were constantly being added to the gene pool, he said, in the form of human waste and runoff.

Epstein lobbied scientists working in fields as varied as oceanography, atmospherics, satellite imagery, plankton biology, and epidemiology to find ways to collaborate, and answer questions about the links between the marine environment and human health. Epstein discovered that many other scientists had already reached the conclusion that changes in global ecology—particularly those caused by warming—were too often working to the advantage of the microbes.

For example, at Yale, where he still ran the Arbovirus Laboratory, Robert Shope was considering the impact of global warming on disease-carrying insects. On the basis of his nearly forty years of arbovirus research, Shope was convinced that even a minor rise in global temperature could expand the territory of two key mosquito species: *Aedes aegypti* and *A. albopictus*. Both species were limited geographically in the 1990s by climate. *A. aegypti* couldn't withstand prolonged exposure to temperatures below 48°F and died after less than an hour of 32°F weather. *A. albopictus* was only slightly heartier in cold climes. As a result, in the Northern Hemisphere *A. aegypti* couldn't live above 35°N latitude, or roughly the levels of Memphis, Tennessee, Tangier, Morocco, and Osaka. *A. albopictus* couldn't survive above 42°N latitude during the 1990s, roughly equivalent to Madrid, Istanbul, Beijing, and Philadelphia.

Shope expected that warming would allow both mosquito species to comfortably move northward, invading population centers such as Tokyo, Rome, and New York. *A. albopictus,* the Asian tiger mosquito, could carry the dengue virus. *A. aegypti* was more worrisome because it carried both dengue and yellow fever; the latter was typically fatal 50 percent of the time. Historical analysis seemed to confirm such a hypothesis, as malaria had shifted geographically over the millennia in accordance with major climate changes.

British experts on insect-borne diseases felt certain that global warming would greatly expand the territory and infectivity ratio of the East African tsetse fly, which carried the trypanosomes responsible for sleeping sickness. The researchers concluded that even a moderate increase—on the order of 1° to 2°C—could result in a higher rate of disease spread because the tsetse flies were known to be more active, to feed at a higher pace, and to process trypanosomes more rapidly at higher ambient temperatures. Thus, each tsetse fly could infect more people daily.

The same principles held true for *Anopheles* mosquitoes and the spread of malaria. In 1993, Uwe Brinkmann, who headed the Harvard Working Group on New and Resurgent Diseases, was trying to figure out ways to predict not only latitude movements of mosquitoes in response to global warming but also their altitude changes. He felt there was an urgent need for research to determine which factors played a greater role in limiting *Anopheles* activities at altitudes above 500 feet: air pressure or cooler temperatures. If the latter was more important, he predicted, malaria could quickly overtake mountainous areas of Zimbabwe, Botswana, Swaziland, Rwanda, Tanzania, Kenya, and other geographically diverse parts of Africa. Further, the disease might with global warming climb its way further up the foothills of the Himalayas, the Sulaiman Range, the Pir Panjal, and other mountainous regions of Asia.

A detailed WHO Task Group report in 1990 offered a broader range of expected disease impacts from global warming. Even a moderate net temperature increase—on the order of 1°C—would alter wind patterns, change levels of relative humidity and rainfall, produce a rise in sea levels, and widen the global extremes between desert regions and areas afflicted with periodic flooding. These conditions would, in turn, radically alter the ecologies of microbes that were carried by insects. Furthermore, expected changes in vegetation patterns could, the WHO Task Group said, radically alter the ecologies of microbe-carrying animals, such as monkeys, rats, mice, and bats, bringing those vectors into closer proximity to *Homo sapiens*.

There was also a strong consensus among immunologists that heightened exposure to ultraviolet light—particularly UV-B radiation—suppressed the human immune response, thus increasing *Homo sapiens*' susceptibility to all microbes. Just as PCBs and other hydrocarbon pollutants were thought to have played a role in increasing microbial susceptibility in marine mammals, so many physicians felt there was ample evidence that air, water, and food pollutants affected the human immune system.

Another feature of global warming would be an increased dependence in wealthier nations on air conditioning. In order to conserve energy, buildings in the industrialized world had specifically been designed to minimize outward and inward air flow. It was much cheaper to heat or cool the same air repeatedly in a sealed room than to pump in fresh air from the outside, alter its temperature, circulate it throughout a structure, and at the same time expel old air. As the numbers of hot days per year increased, necessitating longer periods of reliance upon air conditioning, the economic pressures to recirculate old air repeatedly, to the limits of reasonable oxygen depletion, could be expected. Such practices for winter heat conservation in large office buildings had already been linked to workplace transmission of influenza and common cold viruses. Spread of Legionnaires' Disease and other airborne microbes was expected to increase with global warming.

Even in the absence of serious global warming, energy conservation practices were, for purely economic reasons, spurring architects and developers toward construction of buildings that lacked any openable windows and were sealed so tightly that residents were apt to suffer "sick building syndrome": the result of inhaling formaldehyde, radon, and other chemicals present in the building foundation or structure. Such chemicals posed little threat to human health if diluted in fresh air, but were significant contributors to health problems in residents and employees who inhaled levels that were concentrated in recirculated or thin air. Obviously, a building that was capable of concentrating such trace chemicals in the air breathed by its inhabitants would also serve as an ideal setting for rapid dissemination of *Mycobacterium tuberculosis,* if an individual who suffered from active pulmonary disease was residing or working within the structure.

The human lung, as an ecosphere, was designed to take in 20,000 liters of air each day, or roughly 60 pounds. Its surface was highly variegated, composed of hundreds of millions of tiny branches, at the ends of which were the minute bronchioles that actively absorbed oxygen molecules. The actual surface area of the human lung was, therefore, about 150 square meters, or "about the size of an Olympic tennis court," as Harvard Medical School pulmonary expert Joseph Brain put it.

Less than 0.64 micron, or just under one one-hundred-thousandth of an inch, was all the distance that separated the air environment in the lungs from the human bloodstream.

All a microbe had to do to gain entry to the human bloodstream was get past that 0.64 micron of protection. Viruses accomplished the task by accumulating inside epithelial cells in the airways and creating enough local damage to open up a hole of less than a millionth of an inch in diameter. Some viruses, such as those that caused common colds, were so well adapted to the human lung that they had special proteins on their surfaces which locked on to the epithelial cells. Larger microbes, such as the tuberculosis bacteria, gained entry via the immune system's macrophages. They were specially adapted to recognize and lock on to the large macrophages that were distributed throughout pulmonary tissue. Though it was the job of macrophages to seek out and destroy such invaders, many microbes had adapted ways to fool the cells into ingesting them. Once inside the macrophages, the microbes got a free ride into the blood or the lymphatic system, enabling them to reach destinations all over the human body.

The best way to protect the lungs was to provide them with 20,000 liters per day of fresh, clean, oxygen-rich air. The air flushed out the system.

Dirty air—that which contained pollutant particles, dust, or microbes—assaulted the delicate alveoli and bronchioles, and there was a synergism of action. People who, for example, smoked cigarettes or worked in coal mines were more susceptible to all respiratory infectious diseases: colds, flu, tuberculosis, pneumonia, and bronchitis.

Because of its confined internal atmosphere, the vehicle responsible for the great globalization of humanity—the jet airplane—could be a source of microbial transmission. Everybody on board an airplane shared the same air. It was, therefore, easy for one ailing passenger or crew member to pass a respiratory microbe on to many, if not all, on board. The longer the flight, and the fewer the number of air exchanges in which outside air was flushed through the cabin, the greater the risk.

In 1977, for example, fifty-four passengers were grounded together for three hours while their plane underwent repairs in Alaska. None of the passengers left the aircraft, and to save fuel the air conditioning was switched off. For three hours the fifty-four passengers breathed the same air over and over again. One woman had influenza: over the following week 72 percent of her fellow passengers came down with the flu; genetically identical strains were found in everyone.

Following the worldwide oil crisis of the 1970s, the airlines industry looked for ways to reduce fuel use. An obvious place to start was with air circulation, since it cost a great deal of fuel to draw icy air in from outside the aircraft, adjust its temperature to a comfortable 65°–70°F, and maintain cabin pressure. Prior to 1985 commercial aircraft performed that function every three minutes, which meant most passengers and crew breathed fresh air throughout their flight. But virtually all aircraft built after 1985 were specifically designed to circulate air less frequently; a mix of old and fresh air circulated once every seven minutes, and total flushing of the aircraft could take up to thirty minutes. Flight crews increasingly complained of dizziness, flu, colds, headaches, and nausea.

Studies of aircraft cabins revealed excessive levels of carbon dioxide—up to 50 percent above U.S. legal standards. Air quality for fully booked airliners failed to meet any basic standards for U.S. workplaces.

In 1992 and 1993 the CDC investigated four instances of apparent transmission of tuberculosis aboard aircraft. In one case, a flight attendant passed TB on to twenty-three crew members over the course of several flights.

Similar concerns regarding confined spaces were raised about institutional settings, such as prisons and dormitories, where often excessive numbers of people were co-housed in energy-efficient settings.

In preparation for the June 1992 United Nations Earth Summit in Rio de Janeiro, the World Health Organization reviewed available data on expected health effects of global warming and pollution. WHO concluded that evidence of increased human susceptibility to infectious diseases, due to UV-B immune system damage and pollutant impacts on the lungs and immune system, was compelling. The agency was similarly impressed with estimates of current and projected changes in the ecology of disease vectors, particularly insects.

It wasn't necessary, of course, for the earth to undergo a 1°–5°C temperature shift in order for diseases to emerge. As events since 1960 had demonstrated, other, quite contemporary factors were at play. The ecological relationship between *Homo sapiens* and microbes had been out of balance for a long time.

The "disease cowboys"—scientists like Karl Johnson, Pierre Sureau, Joe McCormick, Peter Piot, and Pat Webb—had long ago witnessed the results of human incursion into new niches or alteration of old niches. Perhaps entomologist E. O. Wilson, when asked, "How many disease-carrying reservoir and vector species await discovery in the earth's rain forests?" best summed up the predicament: "That is unknown and unknowable. The scale of the unknown is simply too vast to even permit speculation."

Thanks to changes in *Homo sapiens'* activities, in the ways in which the human species lived and worked on the planet at the end of the twentieth century, microbes no longer remained confined to remote ecospheres or rare reservoir species: for them, the earth had truly become a Global Village. Between 1950 and 1990 the number of passengers aboard international commercial air flights soared from 2 million to 280 million. Domestic passengers flying within the United States reached 424 million in 1990. Infected human beings were moving rapidly about the planet, and the number of air passengers was expected to double by the year 2000, approaching 600 million on international flights.

Once microbes reached new locales, increasing human population and urbanization ensured that even relatively poorly transmissible microbes faced ever-improving statistical odds of being spread from person to person. The overall density of average numbers of human beings residing on a square mile of land on the earth rose steadily every year. In the United States, even adjusting for the increased land mass of the country over time, density (according to U.S. census figures) rose as follows [see table below].

Though the population was spread unevenly over a country, density trends remained favorable to the microbes. If worst-case projections for human population size came to pass, some regions would have densities in excess of 3,000 people per square mile. At that rate the distinctions between cities, suburbs, and outlying towns would blur and few barriers for person-to-person spread of microbes would remain.

With the passage of time and the increase in travel it was becoming more and more difficult to pinpoint where, exactly, a microbe first emerged. The human immunodeficiency virus was a classic case in point, as it surfaced simultaneously on three continents and spread swiftly around the globe.

Those scientists in the 1990s whose primary focus was viruses believed that the worst scales of disease and death arose from epizootic events: the movement of viruses between species. In such instances, the hosts were usually highly susceptible, as they lacked immunity to the new microbe. Ebola, PDV-2, Marburg, Machupo, Lassa, and Swine Flu

Year	Total Population	Persons per Square Mile
1790	3,929,214	4.5
1820	9,638,453	5.5
1850	23,191,876	7.9
1870	39,818,449	13.4
1890	62,947,714	21.2
1910	91,972,266	31.0
1930	122,775,046	41.2
1950	151,325,798	42.6
1970	203,211,926	57.5
1990	250,410,000	70.3
1992	256,561,239	70.4

were all examples of such apparently sudden emergences into the *Homo sapiens* population.

Rockefeller University's Stephen Morse, who by 1988 was devoting nearly all his professional energies to emerging disease problems, labeled these movements of viruses between host species "viral trafficking." He considered the world's fauna a vast "zoonotic pool," each species carrying within itself an assortment of microbes that might jump across species barriers under the proper circumstances to infect an entirely different type of host.

[...]

Yellow fever, for example, could for decades on end afflict virtually no *Homo sapiens* in a given area because the *Aedes aegypti* mosquitoes were busy feeding on monkeys and marmosets in the jungle. But with changes in either the forest environment or the social behaviors of local *Homo sapiens* mosquito could almost overnight change its feeding patterns and a human epidemic would commence. Such was the case with yellow fever epidemics in Nigeria and Kenya in 1987, 1988, 1990, and 1993.

Tom Monath had seen it happen several times in West Africa, where such simple actions as chopping down a stand of trees and leaving the stumps in place could spawn a yellow fever outbreak. The mosquitoes left their larvae in rainwater that collected in the tree stumps.

[...]

Many insect-borne viruses were thought to have originally been plant microbes that, thousands of millions of years ago, infected insects as they fed on plant nectar. In the 1990s, amid evidence of rising rates of genetic change in many plant microbes, concern was expressed about the possible emergence of new species that might be absorbed by insects. In such a scenario, a microbe that was genuinely new, to which humans had no natural immunity, might quite suddenly emerge. Genetic change in plant microbes was accelerating due to agricultural practices that exerted strong selection pressures on the microbes; to changing geography of plant growth due to international trading of plant seeds and breeding practices; and to the deliberate release of laboratory genetically altered plant viruses that were intended to offer agricultural crops protection against pests.

To minimize use of toxic pesticides, and to prevent incurable viral diseases in plants, scientists in the 1990s were developing ingenious genetic means to protect plants. Using crippled viruses to carry genes that would help vital food crops fend off dangerous pathogens, researchers were breeding plants that could withstand a range of types of infections. There was a catch, however. Studies showed that, in nature, plants such as corn, wheat, and tomatoes were commonly co-infected with up to five different viruses, and those viruses could exchange genetic material. A review of 125 plant strains produced through such laboratory manipulation showed that 3 percent of the time the crippled virus that was used to carry such genes into plant cells could swap genes with other viruses in the plant, producing active, pathogenic—*new*—viral species.

[...]

Theorists were busy trying to determine whether the balances between human immunity and microbial virulence were tipped by any particular identifiable contemporary factors. Nobel laureate Dr. Thomas Weller expressed concern that the ever-increasing numbers of severely immunosuppressed people on the planet posed a real threat for emergence of new disease problems. Cancer patients treated with high doses of chemotherapy or radiation, people infected with HIV, and individuals undergoing transplant operations all represented potential breeding sites for new or mutated microbes. Weller worried about a possible "piggyback" effect, with one microbial population taking advantage of severe immunodeficiencies produced by another microbe or medical treatment.

Another population of immunosuppressed individuals consisted of those suffering from chronic malnutrition. Wherever a significant percentage of the *Homo sapiens* population was starving was likely to be a spawning ground for disease.

Vaccines, where available, protected people against disease, but not against infection. Microbes could enter the body, but even highly virulent organisms found themselves facing an immune system that was primed and ready to mass-produce antibodies. Battles ensued; the invader was vanquished. If a sufficient number of *Homo sapiens* in a given area possessed such immunity it would be possible to essentially eliminate the microbe. Unable to find a *Homo sapiens* host in which it could replicate, the microbial population would nearly disappear. Nearly. In this state, known as herd immunity, humans (or livestock animals) never suffered disease, though they might be infected, unless the necessary level of immunity in the overall population slacked off. For that reason, schoolchildren vaccine campaigns had to reach a critical threshold of successful completion or the unvaccinated children would be a great risk for disease.

Herd immunity faced tough challenges in the age of air travel because individuals who carried microbes to which they were personally immune could fly into geographic areas where herd immunity was extremely low. Under such circumstances, even organisms not generally thought to be particularly virulent could produce devastating epidemics.

The best example of the phenomenon was the estimated 56 million American Indians who succumbed to disease following the arrival of Europeans—and their microbes. That die-off continued 500 years later, into the 1990s, as Old World microbes reached the Xikrin, Surui, and other Amazon Indians.

[...]

[From 1980 to 1989, the incidence of] refugees fleeing natural disasters, wars, famine, or oppression increased by 75 percent every year. By the end of 1992, according to the United Nations, 17.5 million *Homo sapiens* were refugees, most of them living in squalor in the world's poorest countries.

Thirdworldization had set in all over the globe. Millions of abandoned children roamed the streets of the world's largest cities, injecting drugs, practicing prostitution, and living on the most dangerous margins of society. Western European unemployment soared, from less than 3 percent in 1970 to more than 11 percent in 1993, and a sense of hopelessness cast a pall over much of the continent. Civil war in the horribly overcrowded nation of Rwanda broke into inconceivable carnage during the spring of 1994. Serb invasions of Bosnia devolved into little more than slaughter of civilians.

Conservative Harvard University political analyst Samuel P. Huntington opined that the world had entered a stage of conflict that superseded nation-states, economic competition, and ideologies, becoming something far more insidious: cultural conflict. Wars and battles were fought over religion, over historic enmities that in some cases traced back to slights that had transpired between opponents more than 2,000 years ago.

In such a context, it seemed difficult to discuss *E. coli* virulence mutation probabilities. If men in the former Yugoslavia considered multiple acts of gang rape of civilian women justified acts of war, how could there be rational discussion of probabilities of sexual transmission of disease?

Still, the scientists pushed on, determined to remain cool in the face of global disarray, perhaps *because* of the chaos which threatened to abet the microbes. Studies demonstrated the rapid spread of disease among refugees and the emergence of antibiotic-resistant bacteria and drug-resistant parasites in such clusters of humanity. The health risks of famine were carefully tallied.

A concern shared by all public health observers was the shift globally from low-intensity geopolitical nuclear confrontation to high-intensity local conflicts. While the former had posed the threat of thermonuclear war, little actual conflict occurred. With the fall of the Berlin Wall dawned an era of extremely high-intensity conventional and guerrilla conflict which took a tremendous toll on civilian populations: direct losses of life, homelessness, refugee migrations, demolition of basic infrastructures, destruction of hospitals, and, in some cases, pointed deliberate assassinations of health providers.

When such conflicts occurred in developing countries, they created new possibilities for re-emergence of old scourges such as typhus, cholera, tuberculosis, and measles—the classic wartime opportunists. Where sex became an economic component of strife, microbes that could exploit sexual transmission emerged. And along the peripheries of human battle and despair lurked the unexpected. In the flora and fauna of remote ecospheres [where] they resided, human events affording them ever-greater opportunities for jumping from their ancient hosts to the warring *Homo sapiens*.

Song for the Blue Ocean

By Carl Safina

Introduction

Carl Safina's *Song for the Blue Ocean* is perhaps the most beautifully written of all the selections in this reader. It combines good science with inspiring (and often saddening) depictions of the state of our global fisheries. "Sulu," the final chapter, is close to my heart because it concerns the southern Philippines, whence hail my wife and her siblings. The Sulu Sea area is at the heart of the Coral Triangle, the absolute center of marine biodiversity for the planet. This area is also at the intersection of three land-based biodiversity hotspots (the Philippines, Sundaland, and Wallacea) and is home to numerous tribal and ethnolinguistic groups as well; the human diversity equals the biodiversity. Animist tribal groups (called "Lumad" in the Philippines) and so-called Bajaus share the mountains, fields, and reefs with Buddhist/Confucian Chinese, with Catholics from the Visayan archipelago of the Central Philippines, and with Muslims, both the fierce Tausugs and the peaceful Sama. They are farmers, fishermen, shopkeepers, and traders living in a region legendary for its beauty and natural riches, as well as for its lawlessness and corruption. Pirates still roam the coast of North Borneo as they have for centuries. Local politicians often have private militias and run their domains as personal fiefdoms (as Safina so eloquently describes).

How on Earth could biodiversity conservation be possible in such a tragic and troubled region? Safina describes this in great detail. I can only add that this work is ongoing and has spread. The key is this: The locals are not stupid. For centuries they have reaped the bounty of the oceans and forests, and they know what it is for this bounty to fade away due to overpopulation, corruption, shortsightedness, and from the desperate acts of the poor. Successful efforts in this region, as in other parts of the developing world, center themselves on the community. I have seen good men like Filemon Romero and Nur Haroun, as part of the WWF-funded FISH Project, go among their own people, from village to village, exhorting them: Protect this one small area and your fish will come back. Build one guard tower. Find a boat and a gun and a brave man, and empower him to defend your tiny marine reserve, and your fish will come back. During a visit to Tawi-Tawi, a province of the Philippines' Autonomous Region of Muslim Mindanao, I learned that dynamite fishing was down 97% after one year of these heroic efforts. If, as Safina describes, even militiamen can see the benefits of stopping cyanide fishing, there must be some hope for this amazing planet. We have a long way to go, of course. Corrupt officials must be kept from selling off the last of the forests, or mudslides will kill more people, as happened when a storm washed the soil off the denuded mountains and into the Visayan town of Ormoc, killing 8000 people. Ignorance must be challenged. On my last visit, a local leader showed me a Philippine Eagle (the national bird) that he kept as a pet. I gently suggested he contact the Philippine Eagle Foundation; I did not, however, confront my generous hosts when they offered me turtle eggs, a "treat" I hope I never see on a table again. As an outsider, it's important to choose your battles.

I had a three-man bodyguard team when I visited Tawi-Tawi. Safina went to Jolo, a much more dangerous region, and did so without a bodyguard. His (perhaps unwitting) bravery moves me almost as much as the grace of his prose and the power of his science. I hope it also moves you.

I have been to the edge of human existence before, but this will prove different from anything in my experience. For the moment, though, I have no way of knowing what I am getting into. After a sleepless night of harried travel, I am still too wound up to doze as the first tint of dawn outlines the Philippine horizon outside the airplane's window. When I was in Palau, Larry Sharron gave me a copy of a letter to his father, and he told me to read it when I got to the Philippines. Now I pull the envelope from my bag.

Dear Dad,

Well, I'm back from the Philippines now. And what a long, hard trip it's been. After 5 weeks there I was ready to come back to Palau. I'll tell you a little about the trip.

Palawan is the most southwestern large island in the Philippines, and we went there because we felt that the marine life would be different from other places we had been. Our first few days were spent diving from shore. I must admit the invertebrates were interesting. The species we were looking for had no food value, and were abundant. But I couldn't help notice the absence of larger fish. Like in other parts of the Philippines the pressures of the large populations of people have taken their toll.

After about a week we chartered a refitted freighter to take us to a large reef system called Tubbataha, 100 miles offshore, out in the middle of the Sulu Sea. We dove all around Tubbataha for a week. The reefs were beautiful, though some dynamite fishing had occurred. Again, no large fish, and not a single shark. The crew on the ship said the Taiwanese fishing boats had been coming there and caught all the sharks for their fins. I went out several nights looking for various invertebrates. But here, there were very few. And live shells were almost non-existent. The crew of our ship explained to me that the shellers had also been out here in large numbers, but they do not frequent the area anymore. No wonder they don't, they've picked it clean. The seagrass areas in the center of the lagoon were the same—beautiful habitat, no animals. We did observe a large sea turtle leaving her nest one night. Perhaps there is hope for this area. The water was very clear and we did manage to find several new sponges and other invertebrates for our project.

We returned to Palawan for the next phase of our trip, out to San Salvador Island in Zambales Province. What we saw shocked us. The reefs in the area had been decimated and covered with silt. On one entire dive we saw only two fish. Local divers attribute it to dynamite fishing and the use of sodium cyanide to capture fish in and around coral heads. Both these methods have gone on unchecked for the last 20 years. They discovered that cyanide could be used to collect fish when they observed that the runoff from a nearby gold mine was killing fish in the river and eventually the sea. Cyanide is commonly used to remove gold during mining or to dissolve it. It wasn't long before everyone was using this deadly compound to fish with. And for those who don't dive and cannot spray the reef directly with a poison, there is dynamite. When exploded underwater it stuns everything in a 100 ft. diameter, while blowing the reef to bits. Every day that we dove off San Salvador we could feel and hear blasting underwater.

We wondered what they were catching, since we hadn't seen too many fish worth serving up. A trip to the local fish market answered our question. There were no large reef fish for sale, but there were thousands of one-inch baby rabbitfish, dried and salted like potato chips. I guess they just can't wait for them to grow up.

The divers of San Salvador Island are aware of their plight and know the damage they have done. Many are seeking employment in other countries to support their families. At present, if they make 100 pesos ($4.00 U.S.) a day, they can survive. But even their island is fast becoming overcrowded from the growing population. Their fate is gloomy at best. Since I know many of them personally, I feel for them. In an effort to try and return the reefs to their original state, they have set aside the reef on one side of their island as a refuge. No fishing whatsoever. We dove there and found that there may be hope. We observed a few large fish on those reefs. It's a

beginning. They say that the hardest part is policing the area for fishermen from nearby islands. They've set up a small house on the beach in front of the sanctuary and someone is always there watching for illegal fishing activity.

We spent a few days in Manila, working out of the National Museum, and then shipped our stuff off to Washington, D.C., to the National Cancer Institute. We needed dry ice for our samples, and found it at a nearby ice cream factory. But they made us buy one gallon of ice cream for every 2 blocks of dry ice we purchased. We ended up giving the ice cream away to the various museum personnel that had been so helpful. Well that's my story. I hope that wasn't too lengthy or boring. Take care of yourself and stay well.

Lawrence

My gaze drifts out the airplane window at the islands below. Sleeping soundly in the next seat is Dr. Vaughan Pratt, director of the International Marinelife Alliance. At Bob Johannes's suggestion, I had contacted Pratt about his work to end cyanide fishing in the Philippines, and he had invited me to accompany him on his first trip to one end of the world. Situated between the Sulu and Celebes Seas, the island of Mindanao, and Borneo's Malaysian Sabah region, the predominantly Islamic islands of Basilan, Jolo, and Tawitawi and their many out islands form the southern tip of the Philippines—the Sulu Archipelago. There, the country jags and shatters into splinters, geographically and politically.

Though the Philippines were endowed naturally as part of the world's richest marine region, years of abuse have turned them into perhaps the world's most ravaged coastal ruins. By the late 1980s, the condition of a third of the reefs in the Philippines was poor, 40 percent were fair, a quarter good, only 5 percent excellent. Over the long haul, habitat destruction poses more of a problem than overfishing in the tropics. Most effects of overfishing, if corrected, can be reversed in ten or twenty years, if the habitat is there. But coral recovery takes many decades.

Much of the Philippines' coral reef destruction followed forest clear-cutting during Ferdinand Marcos's reign. As tree cover declined from 80 to 20 percent, deforestation caused major soil erosion. Sediment runoff quadrupled. Eroded soil, carried by rivers out to sea settled on the coral reefs, smothering them in places. The fisheries and tourism destroyed by logging had yielded higher total profits than the logging itself, according to the coral reef specialist Don McAllister, who published several technical papers on the Philippines. And the use of explosives and cyanide by fishers destroyed much of the coral. These practices are illegal, which doesn't count for much. It just means that everyone doing it is in violation of the law. A conservatively estimated 330 tons of cyanide per year, year after year, waft over Philippine reefs. Forget the number. Take-home message: It's hit virtually everywhere.

But some of Sulu's remote reefs remain more intact than most, and Sulu is where we are headed. Pratt himself has never been to Sulu. His mission here: to help establish a program training cyanide Fishermen to instead use nets for collecting aquarium fishes.

High-priced angelfishes, emperors, tangs—especially the blue tang—and several species of damselfish and basslets found nowhere else haunt the islands and offshore reefs of the Sulu Sea. Groupers, which also inhabit these areas, often end up in the live-export trade and, eventually, on dinner plates in Hong Kong. Fishers here receive five times as much money for live groupers as for dead, and many have turned away from blast fishing, which kills fish, toward sodium cyanide, which stuns them.

It is not an improvement. I once spoke to a former cyanide fisherman from the village of Nasugbu, south of Manila Bay. He is now a provincial fish examiner and leader of a three-thousand-member group formed to end destructive fishing. He knew the horrors of cyanide firsthand, having known fishers who died underwater when they swam through milky clouds of cyanide they had squirted. He'd also known a fisher who died after eating with cyanide-contaminated hands, and several children who died after eating cyanide-contaminated fish. I asked whether he could catch more fish with cyanide. He answered, "Yes! Yes, of course. You catch plenty in a short time. But when we went back to our fishing areas there were dead corals and no fish. The snails and shellfish we collected for food disappeared. We are our own victims. Now we want to help ourselves."

Vaughan Pratt is trying to help people help themselves. Vigorous and energetic, Pratt is in his early forties, tall, square of jaw, his slightly graying, wavy hair swept back. Pratt first came to the Philippines from the U.S. in the 1970s for graduate studies in veterinary medicine, inspired by a childhood desire to help animals. After completing his doctorate he returned to the States for several years, going into private veterinary practice. "After a while," he explained, "it got obvious that killing fleas on Mr. Wilson's dog was not saving the world." So he got into rehabilitating injured wildlife: animals hit by cars, birds that had hit wires or become tangled in fishing line, geese with broken wings, those sorts of cases. "Soon I realized you can't quite save the world one animal at a time. It's too slow that way." He moved on to entire groups of animals and their habitats. "And what habitat," he says, "is larger than the oceans?"

If that seems idealistic, it is. Idyllic, it is not. Pratt has bitten off a difficult, dangerous chunk of work. He now lives in Manila with his Philippine wife and their three children. To me, ending the use of cyanide in Philippine fishing seems nearly impossible. For one thing, the chemical is widely available. As Pratt once said, "In addition to fishing, cyanide is used in pesticides and to fumigate. So we have cyanided vegetables, cyanided rice, cyanided fish, and now they've legalized the death penalty, so maybe we should donate all the cyanide we confiscate to the gas chambers." More to the point, cyanide fishing is already illegal, and has been since 1974. It is done out of sight. It is ingrained. Cyanide was introduced in the 1950s, so three generations of fishermen have a tradition of using it. But Pratt says, "Our slogan is: 'Toward a cyanide-free fishing tradition.'"

That slogan is his guiding vision. For a visionary, Pratt is extraordinarily practical. His strategy for achieving the impossible involves a comprehensive plan of action so simple and pragmatic I think it could work: 1) Teach fishers how to use nets and hook-and-line setups. (Nets and hooks might seem pretty basic to fishing, but many Filipino fishers who use explosives or cyanide have never learned to use them.) 2) Spot-check all live-fish shipments for illegal cyanide, using a chemical technique he helped develop. 3) Make authorities ready—and willing—to seize shipments that test positive.

Envisioning such a plan of action is an impressive first step. Implementing the vision will take a lot more than imagination. It will take networking, establishing credibility, securing agreements with government, obtaining funding, building and staffing testing labs, and being reachable by phone at all hours to catch stuff when it falls.

Vaughan's group now has five offices, with four cyanide-detection labs and fifty staff members on the payroll. ("On the payroll" is more accurate than "paid staff," since many of them go unpaid at times when government contracts get hung up or private contributions from the U.S. lag.) The whole operation is running on a shoestring of about $200,000. Salaries are very low by American standards; most workers earn $200 to $300 per month. Vaughan himself relies on income from his wife's hair-bow manufactory, which occupies a room adjacent to their kitchen.

Vaughan's commitment to his work stems from his belief that destructive fishing contributes to human misery, displacing thousands of former fishing families to urban slums. "The reason we're here," he told me, "is that exporters were exploiting people. The cyanide trade is like drug pushing. You hook the fishermen and then you sell it to them. As the reefs die, the exporters get rich and the people get set up to go hungry. The corals are food factories. The fish supply for three hundred thousand Filipinos is already insufficient. Malnutrition in children ranges up to seventy-five percent in places." Vaughan's commitment is shared by his staff. Never have I seen an office where so many of the staff cheerfully stay so late and the phone so frequently rings on Sunday. I have seen Vaughan on his hands and knees at nine P.M., cleaning the floor with a rag, while the chemists patiently awaited the delayed arrival of some grouper samples that had to be tested by morning. They sometimes conduct analyses late into the night, napping on the couches while the tests run.

But commitment, patience, and sacrifice are not enough. In this country it takes unusual courage. Tangling with illegal big-money fishing can be more hazardous to your health than handling cyanide. Recently the cyanide-detection labs have been generating about fifty court cases a year. They have also been generating death threats.

Pratt once told me, "It's been dangerous for us because in places we've eliminated the middleman who bought the fish and supplied the cyanide. I don't want a lot of news coverage. I don't want to let the bad guys know what I'm doing or what I look like. It's too easy to get blown away here. People who got kickbacks to look the other way are not too happy with us. It bothers my wife when government people come to my house at five in the morning with guns to 'investigate' us. But what are you going to do, quit? What would that accomplish?"

The net-training project Pratt is taking me to see, "Net and Let Live," is a joint endeavor of the Philippine Ecology Foundation—a local group—and Pratt's International Marinelife Alliance, in collaboration with the local and national governments. The program is brand-new; the first training session was concluded just last month. Training includes catching fish without poison; techniques for handling, decompression, and shipping with maximum survival; and safe diving using compressed air.

I am flipping through a report by one of the local trainers, which says,

> The approach is through community enterprise for the greater benefit of the fisher folk in our country's most depressed areas, particularly in remote and farflung island reef systems, in which our government is unable to monitor or enforce rampant illegal fishing operations because of the lack of resources. These depressed areas have been deprived of much economic and social services in the past.

The thick mud outside the airport in the city of Zamboanga (population several hundred thousand) keeps forming heavy balls on my shoes, and I keep stopping to scrape it off. It is dawn on the island of Mindanao—our stop enroute from Manila to Jolo—and the streets, such as they are, are beginning to swell with pedestrians, bikes, taxis, and peddlers. We are considering breakfast among the peddlers' carts. Vaughan says, "As you can see, here everyone eats fish, and if there are none, people will die." That seems a bit melodramatic. Having had an hour of sleep, I am groggy and not fully attentive. Vaughan tells me to stay alert and not to get separated. "This is not the safest place in the world."

After breakfast, we board a midmorning flight in a small plane to the town of Jolo on the island of the same name, in the Sulu group. I sit back and read a newspaper account of the murder of a member of a local conservation group in Zambales. That's near where Larry Sharron had been when he wrote to his father. Shot several times in front of his family while having supper at his residence, the victim was vice president of an organization of civilian volunteers, deputized to protect a local bay from illegal fishing activities. After apprehending a number of fishermen engaged in cyanide and blast fishing, he had received threats of physical harm from these groups. Wow, this is scary.

Pratt says, "Yeah, but the country as a whole has improved a lot."

Another article says Islamic terrorists are targeting American and Saudi Arabian nationals in this area. This jogs my memory. When I was in Palau, Devon Ludwig had told me, "In Sulu in the Philippines, the Islamic extremists look upon Americans as potential hostages."

Vaughan says, "They're always saying that Americans are targets. Let 'em try, I haven't heard of any real trouble for a while." Unfazed, he points to a computer ad on the same page and shows me the hardware he's saving up for.

The door of the plane opens, I step out and the sun leans forward to plant a kiss on my forehead. We walk across a dusty runway to the tin-roofed shed that functions as the terminal. It is hot inside, too, but at least it's shady. Our host is not here to meet us as planned, so we sit on the wooden benches along with airport loiterers of uncertain intent, watching chickens pecking at the dirt in the glaring light outside the doorway.

The place seems sleepy. Vaughan says, "This region is the most peaceful it's been in twenty years. I haven't heard of much trouble since a missionary was kidnapped a year and a half ago. Back in 1974, the Moro National Liberation Front, battling government forces, burned this town to the ground."

What?

"Oh, yeah. There are three active political movements: the MNLF—which wants more autonomy, more local control—the National Government, and

Abu Sayyaf." Vaughan suddenly looks up above my head, saying, "Hello, Pete!" We rise to meet our host.

"Ah, yes. Abu Sayyaf! Hello, Dr. Pratt!" He greets Vaughan, then takes my hand, saying, "Extremists. Terrorists. Fundamentalist Muslim separatists who espouse their cause through kidnapping and killing." Then, to Vaughan, "They have again become active, Dr. Pratt!" Perfecto (Pete) Pascua has a turban wrapped around his head and is wearing military fatigues stenciled "Philippine Ecology Foundation," the name of the local group of which he is president. Several men stand behind him—all armed with rifles except one. The exception is wearing a white T-shirt and black pants. "Dr. Pratt, Dr. Safina, this is Bishop DeJesus."

The bishop, a middle-aged Filipino with a hint of Chinese features, extends his hand, saying cheerfully, "Congratulations, you are still free! We were a bit concerned. Two Americans and two Australians were kidnapped while they were waiting for the plane here a few weeks ago."

It takes me a moment to process this. Before I can ask Vaughan why the hell he has brought me, blundering ignorantly, into a war zone without briefing me, Perfecto changes the subject, saying breezily, "The Catholics and we Moslems are working together."

"Yes," the smiling Bishop DeJesus says. "During mass, I mention that we must take care of nature so it can take care of us. I have been here thirty years, and I have seen how cyanide and dynamite fishing are disastrous for our people. So when Pete told me about the arrival of the training program, it was good news. We Catholics are only four percent of the population, but I have offered our classrooms."

We walk outside, where a pickup truck bristling with guns awaits us. I am not used to this, *at all.*

Vaughan shrugs and says, "It's a war to save the oceans. If the rest of the environmental community only knew what's happening—"

"If *I* only knew," I interrupt. Simultaneously apprehensive and annoyed at being unprepped, I throw Vaughan a dirty look.

"You're from New York. It's probably more dangerous at home."

"I'm from *outside* New York."

We climb into the truck and head toward Jolo town, population one hundred thousand. Sunshine gives way to drizzle, forming puddles, making mud. Snack stands, thatch-walled tin-roofed huts, bougainvillea flowers, turkeys, cows, goats, chickens, and staring kids line the roads, many of which are unpaved.

As we bounce past small stands of sugarcane, coconuts, bananas, papayas, and unusual forms of eggplants, the bishop explains to me clearly and simply, "If you are an outsider here, you are subject to kidnapping."

Nice.

He tells us several Belgians building a hospital were kidnapped, as was an American priest while saying mass a few months ago.

I'm feeling this is just a tad too late to be getting clued in on this.

The bishop says gravely, "Somewhere along the line, someone must take the risk and be brave, and end the vicious cycle of violence."

I'm from the suburbs. I can't imagine what that kind of bravery would even look like—much less what it would feel like. I ask, "How might someone possibly do that?"

Father DeJesus, looking a bit surprised at my asking, answers, "Well—what you are doing here: to train the people about sustainable fishing, because the problem is that the people are poor. If they can have a more secure income and future, the problems will begin to go away. Of course, some of the military who receive bribes from the cyanide suppliers will be displeased with what you have been doing, but risk is part of progress."

Whoa! I am thinking, "Hey, *I* haven't been *doing* anything. You're confusing me with heroes. Don't make *that* mistake! I'm just visiting. I don't know anything about this. I'm not part of this. If I had known, I wouldn't have come here. My house is on the other side of the world. I'm not from around here. I just came to watch the experts and maybe learn something—that's *all*. Don't you or anybody get the impression I'm with these guys. I'm going home in a few days. I'm interested in fish, not politics."

And I can almost hear his reply. He'd say, "Fishing, politics—it's all the same here."

Clearly, I *am* involved just by being here, like it or not. And I don't. Not one bit. Adventure is fine, but this is too rich for my blood. I have a cold feeling. I whisper tersely to Pratt, "Did you *know* about this? You said this place has been peaceful for years."

"I didn't hear about *all* those kidnappings." Vaughan says. "They must have kept them quiet."

"I hope they keep *our* kidnapping quiet, because if I hear about it, I will be upset."

The bishop says to Vaughan, "They know that if they bother you they will be tangling with the congressman's family forces. That's the word that has gone out."

I say, "What does *that* mean: 'family forces'?"

Vaughan explains, "The local congressman—the guy we are headed to see—is also a warlord. His family is one of nearly two hundred Philippine war-lord families listed in a recent newspaper article."

"A warlord. Meaning *what?*"

"He has raised his own private army. That's who all these guys are."

"And all this armed escort is to protect us against whom?" I inquire. I am beginning to feel I have a right to know.

"Abu Sayyaf," Perfecto says. "Or anybody that would think of harming us. But not necessarily the government or the MNLF. The MNLF is very well

aware of this project, but they like what they have heard."

We stop to drop the bishop off. His bodyguard goes with him in the rain.

None of us can know that in a few days the newspaper will report that Abu Sayyaf has swept into a town north of Zamboanga, mounting the region's largest attack since Jolo was burned in the mid-1970s, killing dozens, including a police chief and an army commander, and wounding nearly a hundred people while robbing banks, setting fires, and turning the town into a battle zone blanketed with thick smoke. The entire four blocks of the town's commercial center will be razed.

A heavily guarded driveway into a walled prop-erty marks the home of the Honorable Bensaudi O. Tulawie, member of the Philippine House of Representatives from the Sulu District. The con-gressman has invited the Cyanide Reform Program here—the project Pratt and Perfecto's people are collaborating on. Several additional people, wear-ing Philippine Ecology Foundation flak jackets and "Net and Let Live" T-shirts, are present, awaiting us. We are led to a spacious veranda outside Tulawie's residence, a large, open-sided, tile-floored area with a big tiled roof over it. Very gracious looking. An adjacent global communications shop features a large satellite dish. Pigeons coo softly from a loft beside the house.

Mr. Tulawie himself is seated on a wicker couch on the veranda when we arrive. He rises to meet us. He is wearing jeans, a T-shirt advertising sneak-ers, and Western-style sandals. At age thirty-four, Tulawie is one of the youngest congressmen ever elected in the Philippines. He is serving his first term, but the look of his surrounding army does not invite the label "freshman." Fatigue-clad gunmen carrying automatic rifles stand casually or idly pace the veranda's perimeter. They seem bored. An auto-matic rifle laid carelessly across a bench is pointed directly at the congressman's back.

We are invited to sit. Tea and cookies are brought. The mood seems stiff. A moment of pregnant silence hangs over us. Searching for something to say, and thinking a little flattery can't hurt, I try, "You look very young to be a congressman."

"I'm old enough," he says tersely.

Feeling awkward, my eyes wander to a portrait hanging on one of the pillars. It is a photograph of Colonel Muammar el-Qaddafi—Libyan "Strongman," international terrorist.

Seeing my eyes pause, the congressman says, "When I was young I looked up to him. But now he has mellowed."

"Oh." Mellowed. Of course.

He turns to Vaughan, which takes the pressure off me. "So, Dr. Pratt," he chuckles just as I am raising my teacup to my lips, "maybe to get the U.S. government to fund our project, we should kidnap Dr. Safina?"

Everyone is smiling. Everyone *else*.

What does Miss Manners advise in situations where your teatime host turns out to be a warlord and is joking about kidnapping you? I can't remember. I believe I'll have a cookie, while that remains possible.

Tulawie tells us that he has recently set up a pharmacy with free medicine for people holding prescriptions. Good idea. I am going to need some muscle relaxants and anti-anxiety drugs pretty soon.

The congressman then launches into a short, semiformal speech. "In five years, we want the Philippines to be the number-one model in the world for marine resources. We will have marine parks and reserves."

This is too weird. For the first time ever, the term "environmental terrorist" makes some sense to me.

Tulawie continues, "Countries in the region are distorting their resources with cyanide fishing. We are in the last virgin islands in our country and we want to preserve them: our reefs and corals and other creatures underwater. Our local people will make use of this project for their own sustenance. Cyanide does not only kill marine resources, it could contaminate our own bodies. Now we need training to divert this into good things. Instead of carrying around firearms, loitering, and plotting to kidnap or murder, our people will be making a living in a nice way."

I especially like that last part.

Sensing that our audience with the congressman is over, we all rise to exchange farewells. The congressman says warmly, "You will be staying overnight in a wonderful place." As we are departing he adds, "Don't worry, be happy. Everything is normal."

I want to believe him, Vaughan and I are shown to a capped pickup truck where a driver is waiting, and we climb into the back. Perfecto will be coming later. We will be driven across the island of Jolo to a small village, from which we will be ferried to the boat that functions as headquarters for the fish-netting training program. In an open vehicle ahead sit six gunmen corsetted with cartridge belts, fingering automatic rifles: our escort. Vaughan says sarcastically, "I'm glad to see we are keeping a low profile."

As we bounce along the heavily potholed dirt roads, I notice that several of the wagging guns in the lead vehicle are pointing directly at our windshield. Each time the escort vehicle hits a hole hard, I brace myself for the possibility that a gun might go off and one of those bullets might poke a space through my brain.

We pass three schools (nice to see a government that values education, anyway) and one absolutely huge lone remaining rain-forest tree towering over what is now a large banana grove. I wonder why that one tree was spared. Many of the houses we see are on stilts, suggesting that flooding might be a problem here during typhoons. And many of the people we see occasionally walking along the road or passing in trucks are carrying guns. I say something to

Vaughan about all the guns. Our driver laughingly informs us, "During the elections time, there is a ban on carrying guns. Elections time is now."

Vaughan says to me, "Don't worry, nobody ever gets shot here accidentally." Vaughan tells me the Moro National Liberation Front is active right in this area. He says, "Peace talks are ongoing. They want an autonomous Muslim region."

Autonomy is looking pretty reasonable right now. I say, let 'em have it if it means that much to them. Because, you know what? I'd like to keep mine.

Clouds sweep across forested slopes of distant mountains. We travel on. After penetrating deeper into Jolo Island for about an hour, we turn off the "main" road onto an even smaller one. But this road is better, with a smoother surface of day.

The sky is thick with clouds, vapor, and passing rain, and for the next few miles we travel through still jungle as though into the very heart of darkness: an implacable hostility, brooding over an inscrutable intention.

We have seen only an occasional dwelling after turning onto this road. The few houses are mere huts constructed entirely of bamboo. Vaughan says a little ominously, "I haven't been in an area *this* remote in a long time." Then he adds brightly, "That's why it's good to work here: You can bring something new."

A drizzle begins falling—pretty steadily—and the road that seemed so smooth acquires a slick, slippery coating as the wet clay turns squishy. We weave and career down declivities, occasionally spinning our wheels when the need to climb presents itself.

After a while, this track narrows, becoming potholed and wash hoarded as well. The flesh on our driver's thick neck shakes as we vibrate along. The wetter the road gets, the more difficulty he is having preventing the vehicle from sliding and skidding. Our armed escort vehicle, appropriately equipped with four-wheel drive, is doing better, pulling farther ahead, occasionally disappearing from view around curves. Losing visual contact with our escort makes our driver look even more worried. He seems to be getting nervous, and he has begun sweating.

Our driver mentions something I don't quite get, something about cosmetics people. This is not exactly the place I'd expect to see the Avon lady, and I ask Vaughan what the driver is referring to. Vaughan says, "He thinks this is near the place where the four cosmetics company people—seaweed buyers were kidnapped. They were held hostage for about three weeks. One was killed but the others escaped."

My humor is altogether gone now. I am scared. This is a bad place. I don't know why I am here. I fall silent, wishing—hoping—for this trip to be over.

Suddenly, the narrow road pinches sharply as it dips into a hollow, at the bottom of which a stream runs through a couple of culverts. Beyond the scream crossing, the road rises steeply out. Our escort vehicle is already climbing the other side, but with some difficulty. We begin sliding down the slope into the dip, fishtailing through the clay, our driver spinning the steering wheel left and right, trying to keep the truck from slipping off the side of the road and flopping over into the stream.

At the bottom of the decline, we hit thick mud.

While our wheels spin hopelessly, our escort up ahead is about to disappear from view. Our driver leans on his horn for about four seconds to get their attention.

I wish he hadn't done that. They don't seem to hear us—or have decided to ignore us—and they drive away, never to reappear.

Vaughan says nervously, "Not a great place to get stuck."

For a newcomer, I must be catching on very quickly, because I already know this.

We all get out. The tires are in mud up to the bottom rims. I am in to my ankles. I notice for the first time that our driver has a handgun tucked into his belt—despite this being elections time.

Vaughan says agitatedly, "I can't believe we're standing out here in the jungle. The whole idea was: Come quickly, go to the island quickly, unobtrusively. Now everybody knows we're here. If my wife knew about this—for a few moments, no one says a word. The shallow stream, about six feet below, is gurgling under the road and running off through a quiet forest—a forest silent but for the white noise of raindrop upon leaf times infinity.

The truck itself is in a very tricky position, angled sideways and only a few inches away from sliding off the road and into the stream, where it would lie half-submerged on its side, with no hope of recovery by us. The first order of business, then, is to straighten the truck. We line up on one side of it. For the congressman, and the people we are scheduled to meet later today, I am wearing the best clothes I have with me: new shirt, new sneakers, new pants.

One! Two! Three! On the first push, the spinning wheels splatter me with mud. Three pushes later, mud is clotted to our feet. Our legs have achieved full encrustation. It is raining on and off. We are wet and hot. A few mosquitoes are annoying us. But at least the truck is straight, no longer in imminent danger of slip-sliding away. This will have to pass for progress.

We chock the wheels with rocks from the stream bank. With hope springing eternal, and fear being the great motivator that it is, we break pieces of bamboo and palms and woody branches and shove them under the wheels. Pratt and I rock the vehicle to inch it forward, while the driver splatters us with more mud. The truck wants to tail out sideways, and we shove with everything we've got, keeping it straight. I slip and fall and come up chocolate-coated. Pieces of bamboo fly as we grunt and shove the vehicle three feet forward until the wheels begin drilling holes in the mud and the vehicle starts rapidly sinking.

"Whoa! Stop! That's enough!" We explain to our driver that the trick is: Don't give too much gas. No spinning the wheels—O.K.? He hasn't done the rock 'n' push routine before. Vaughan and I grew up in places with lots of snow, and lots of sandy beaches to get pickups stuck in, and rockin' 'n' pushin' makes us feel like teenagers again. All right. Chock those wheels. Move the bamboo and everything. Let's not lose any ground here. Let's keep it going. We'll get there, little bit at a time. O.K.—once again. Ready?

The first barefoot gunman appears like the first shark in Hemingway's *The Old Man and the Sea*. His mere arrival, a bad event in itself, is a sign that things are likely to get worse before they get better. If they do get better. Padding catlike, silent, he regards us poker-faced. I look to our sweating driver the way a child looks at you when they've fallen, to see by your reaction whether they are hurt. He is clearly alarmed. He is standing on the other side of the truck's hood, and the gunman cannot see that his hand is on his pistol.

The gunman is holding his rifle across his chest. His demeanor seems calm. He is assessing things. He looks to be about forty years old. Lord knows what he is thinking, or why he is here.

Vaughan is behind me. We are all just facing the man, and he is facing us. I am trying to keep a grip on my nerves. In New York, they say it's worse if you act scared. I don't know about here. I have no idea.

Three teenagers about fifteen years old, two armed with rifles, the third carrying a long, crude, homemade pistol, come filing down the slope. Their brows are knit over wide eyes. I don't like the looks of this *at all*. They seem tense and unsure of themselves, looking to one another as they advance, as though this is a game of chicken. These armed kids are in their prime fanatic years, and this could be their chance to prove themselves. When really bad things happen, this is how they start. My knees turn to rubber. We are outnumbered. We are outgunned. Our "bodyguards" are God-knows-where. for all I know they might have sent these guys. They might be drinking up the payoff right now, chuckling as they chug. Who knows? I'm not from around here. I don't know how any of this works. I'm understanding less and less as the day wears on.

Our driver—bug-eyed, beaded with sweat, tense as a cocked trigger—seems on the verge of freaking. None of the gunmen can see his pistol from where they are standing, and if he decides to start shooting we are all dead. Especially me. I'm way out in the open, standing closest to the strangers. I'm the front-liner. We'll end up facedown in the mud, in the warm rain.

But the thing is, no one has made a move either hostile or friendly. Maybe they are still deciding what to think of us. Maybe just scared of the consequences of mistreating us, not knowing who we are or who might seek retribution. Maybe they know who we are and they like what they have heard about why we are here. Maybe they fear the congressman's army, or think we are all armed, and that is making them hesitate.

In the suspended-animation uncertainty over what is about to happen, I'm trying to think of something that might possibly lower the tension. I decide to smile submissively, wave hello. By my motions I indicate the stuck vehicle, tentatively inviting them to join us in freeing it. My legs are jelly.

With deadpan faces and inscrutable intentions, they decide to help, slinging the guns across their backs. That's one giant step for mankind. We work for a few ineffective minutes while I am calming down. Things seem O.K. but the driver is nerve-racked and fumble-fingered. I am thinking we'd better get out of here while the getting is good. Then somebody else appears, then more villagers, this time unarmed; and one person has a rope. Soon there are about a dozen of us in all, trying to pull and push the truck up the incline for a hot, wet, sweaty, muddy hour.

We finally get the truck up to a reasonably secure resting spot. I assume we will now continue, but our driver, seemingly at his wit's end, turns the truck right around, and it slides immediately all the way back down to where we just came from, getting stuck again in the same place, over the stream. There are now about fifteen people and one rope, with which we all pull the truck up the muddy hill toward the direction from which we had originally come.

Once there, we smile politely, waving a fond farewell. We get in calmly, and the driver floors it, zooming the hell out of there, as fast as the wheels will whirl.

The whole thing—breakdown to fragile freedom—has taken almost two hours. My question is: What now?

Our driver says nothing as he takes us up another road, which ends near a channel in some tall mangroves. The tide is out, and under the trees we are looking at a lot of mud and a little bit of shallow water.

Vaughan asks the driver, "What's the plan?"

Driver smiles. "Plan," he says. He chuckles to himself. "Plan."

We sit in silence, a young girl in a blue sweatshirt with BROOKLYN written on it appears, followed by her little sister, who is wearing a pinkish, floral-print dress. Unlike me, they are scrubbed clean and spotless. They watch us with the large eyes of children, from what they seem to consider a safe distance.

There is an old wooden swing on the grass nearby. I get out to sit on it, and I make faces at the children. Vaughan soon follows. A man with his head wrapped in a turban and carrying—naturally—a rifle on his shoulder and cartridges on his waist walks up to the driver's window.

For the humor value, Pratt repeats something he said earlier. "Don't worry, no one gets shot accidentally here."

This time I won't get upset about it until somebody starts shooting. But this time, also, the driver is perfectly calm. We must be in a relatively safe part of the island. They confer. "O.K.," the driver says to us. "A boat will come here. We will wait."

And so we simply sit on the swing, working the caked mud off our shoes, our socks, our pants. Vaughan's face is streaked with mud, meaning mine must be too.

A mere hour and a half melts by, during which no one gets shot, threatened, kidnapped, intimidated, or even particularly frightened, before a very narrow outrigger canoe with a flop-flop engine arrives. We take our packs and step through the mud out to where there is enough water to float the canoe. Muddy water pours off our pants as we climb in, and we sit in the puddle because there are no seats—not that I was expecting seats.

My nerves are returning to normal; I am again feeling more like a visiting scientist than a terrified, would-be hostage. A strongly running incoming tide is flooding the mangrove forest as our boatman takes us through labyrinthine channels among the biggest mangroves I have ever seen. Thousands of fruit bats hang like eggplants in the exposed branches of the tallest mangroves. Among the most productive seafood nurseries in the world, mangroves shelter and feed trillions of juvenile fishes and a large array of other wildlife, and this forest growing out of the saltwater is just gorgeous.

In a few minutes, though, we come to some areas where the mangroves have been cut down and cleared over several acres. By way of explanation our boatman says, "Shreemp." Shrimp ponds are being dug here, not to feed locals but for export to rich countries. The ponds destroy the mangrove nursery areas whose fish feed the villagers. It seems a shame to see them going in here, liquidating the intertidal forest. Here I go again: The very moment I can stop worrying about imminent danger to my life, I resume worrying about conservation of nature and about social injustice.

Only about a fourth to a tenth of the Philippines original mangrove forests remain, depending on which of several estimates is correct. Mangroves have been gouged for aquaculture ponds, cut for wood, and cleared for housing, and not just here in the Philippines. About half of the world's original mangrove forests are gone. In Indonesia, only 20 percent of the original mangrove forests remain and the government has described the country's coral reefs and mangrove forests as being in critical condition, needing immediate rehabilitation; in 1995 the minister of the environment called on local people to begin planting mangrove trees and restoring the natural coral habitats.

We emerge from the overarching mangroves into a wide bay of islands under a light sky with high clouds. Behind us, behind the mangroves, Jolo Island rises into green hills of broken forest. Before us, palm-fringed dollops of sand and some large, flat islands line the horizon. Here and there, thatch-roofed bamboo houses stand on stilts in the open water. People live here in the bay itself, and we are motoring through a "village" of about two dozen such homes right now. A few families are standing outside their doorways, smiling and waving at us. Nearby, over an area of perhaps ten acres, the netlike contraptions from which these people make their living, growing seaweed, hang in the water from buoys at the surface. People in canoes are tending and harvesting the crop.

Vaughan explains that seaweed exports are worth hundreds of millions of dollars in the Philippines. He says this district is reputed to be among the best in the world for the quality of its seaweed, which is used in agar-based products ranging from cosmetics to ice cream. The stuff these people are growing finds its way into markets all around the globe, and into our households and foods.

Up ahead is the Cyanide Reform Program's mobile headquarters—a big, old, yellow wooden boat of about sixty feet, a floating dinosaur. Short rope ladders are lowered for us as we come alongside. Quite a few people line the rail, looking down at us. I am handing up my gear when I see that Perfecto is already aboard. He asks, "What happened to *you* guys?"

"Stuck in mud," I say.

The boat's bare, planked deck, aft of the pilot-house, is shaded by a wooden roof. The big diesel is sunken into a large, encaged colour in the deck in sort of an exposed "engine room." The boat is a confusion of gear, permeated with the smell of diesel oil: diving masks, weight belts, plastic buckets, knives, snorkels, foam boxes, cardboard boxes, assorted clothing, an air compressor, kerosene lanterns, chairs, coiled breathing hoses, cots, a fire ax, rifles, cartridge belts (there is enough ammunition in view to fight a prolonged engagement with a medium-sized country—such as the Philippines), and about a million other things blocked from view by the dozen and a half or so people milling about. The people themselves include local divers, trainers with the Ecology Foundation, bodyguards, a captain, a mechanic, and a cook. Most are men, but one of the Ecology Foundation people and the cook are women.

The perspiring cook, shy and smiling, is stoking coals smoldering on the rear deck in a pit made from the bottom third of a fifty-five-gallon oil drum. She is a flat-nosed, red-skinned woman in a worn, loose, gray dress. Except for a few stray curls spilling onto her sweaty forehead, her thick black hair is pulled back tightly. She has been squatting at a small pile of fresh fish, preparing them for dinner. Her hands and bare feet are coated with slime, scales, gurry, and blood. She looks about thirty. Vaughan says she is probably seventeen or eighteen.

Perfecto is anxious to show us several fish caught today with nets for the aquarium trade. All of us think the nets have the potential to be these people's salvation. In one study, only about 30 percent of cyanided aquarium fish collected in the villages were still alive by the time they got to Manila, while the survival of net-caught fish was 95 to 98 percent. From 1960 to the late 1980s, the Philippines was the number-one supplier of decorative fishes in the world market. High mortality rates from exposure to cyanide, however, had given Philippine fish a bad name, allowing newcomers like Indonesia and Sri Lanka to corner a huge chunk of the market. In the early 1990s, the Philippine market share was in free fall, going from over $6 million dollars in 1990 to around $1 million in 1992.

I peer into the plastic buckets at a small variety of creatures bright with life: bumble bee–colored anemone fishes, lemon-yellow tangs, and highly valued clown triggerfish with their egg-yolk-colored "lipstick" and big white-polka-dots on black bellies. One fish in particular arrests my attention: a palm-sized fish upon whose indigo background is emblazoned a psychedelic thumbprint of alternating white and electric-blue asymmetrical, concentric rings—an astonishingly beautiful, almost impossible creature. Perfecto says with proud excitement, "This *imperator* is top of the line." It is the emperor angel in its juvenile colors, a fish commanding very high prices. There may be more of them left in Sulu than anywhere else.

As I am bowed over the bucket, transfixed by the emperor's Technicolor clothes, someone rests his fingers gently on my shoulder. "Our only problem is," a voice says, "we are racing against time." I turn and straighten up, looking into the beseeching face of a fuzzy-bearded man with a knit cap, pursed lips, thick glasses, and a slightly crossed eye. His name is Nazhar Muayudi. In an exceptionally soft voice he continues, "You have come to a place where we allow no one. We can control the area, but not the environment. We give an idea—the idea of how to care for the environment. I am a local people here. Most of us have not gone to school. Government programs seldom reach to an area like this. We are working seriously. We believe this program is very fruitful to elevate the livelihood of our people. Before, our people are engaged in cyanide fishing. But in the last few months, many people come and want to join us. Because they are already convinced cyanide is very destructive to the environment. They have already experienced how the fish die. And, of course, they have seen the corals die. We don't want to abandon the environment. All the people—not only our generation—will benefit."

I would not have thought the people in a place this remote could feel so affected. But I am seeing again that—as Johannes kept trying to emphasize, and as I saw in the Pacific Northwest—the key to survival is enlightened local control of natural resource use. And, again, this seems especially true when the market is distant and insensitive to the local depletions it can cause.

In most reef areas, though, the force of market demand reigns, unchecked by local control, and the social and economic effects are serious. Don McAllister calculated that 40 percent of the Philippines' fish yield was lost due to degraded habitat. This corresponds to depriving 3,000,000 Filipinos of seafood. It also corresponds to destroying 130,000 fishing jobs. If each fisher's employment helps support four other family members, then about 650,000 persons are *directly* affected by the loss of employment. Considering the commerce that might have flowed from those lost fish, it seems reasonable to estimate that loss of employment caused by fishery depletion affects more than 1,000,000 persons in the Philippines alone.

There was another estimate, too: that by the late 1980s Philippine fishermen's daily catches had dropped from fifteen or twenty kilos per day down to just one to four kilos, a loss of between roughly 75 and 95 percent. More and more small-scale fishermen were catching only enough fish to feed their families and not enough to sell.

Fish is the most important food item in the Philippines after rice. Seafood contributes about half the animal protein consumed by Filipinos. It provides important vitamins and minerals to persons dependent on rice.

A former minister of social services and development wrote that 75 percent of Filipino children under four years of age suffer from malnutrition. Undernourished children do not reach their human potential and are more prone to physical and mental illnesses. Unemployed and improperly fed people are more prone to antisocial behavior. McAllister challenges us to consider the greater implications: How much of the Philippines' large and expensive standing army is required because of human misery and unrest stemming from environmental destruction? What is the cost of the burden unemployment and malnutrition places on health-care services and correctional institutions? As I've seen and heard today, none of these questions is unimportant, none of these issues exaggerated.

The big exposed diesel engine fires up with a tremendous clanking roar. Nazhar says, "Now we are going to the village on Dong-dong Island. You are very lucky. Few outsiders have seen how beautiful it is here. This is the last frontier in the Philippines."

Our yellow dinosaur chugs slowly along in calm water and humid air under a close, overcast sky. Islands and islets ring the horizon, and only a few far-flung huts and villages are visible in an immensity of water and a wide stretch of green coast. This indeed is very much the frontier, the outskirts, the edge, the end of the reach of settlement and development. Yet even here, the people say they are racing against time.

I am seated along the tail, watching the shoreline crawling slowly past. Nazhar pulls up a chair and sits in front of me, knee-to-knee, placing his hands lightly upon my shoulders and looking into my eyes. I feel uncomfortable. He continues softly, "We believe this is our treasure. Our products are ornamental fish, food fish, sea cucumbers, abalone, shells, sharks, sea horses, seaweed. Of all these products we are very much aware. We are looking for a livelihood so that we can eat. But we must not waste the coral. We must not waste the fish. Our religion teaches: You drink, you eat, but do *not* waste. This is the *most* fundamental commandment. Wasting is an evil. It is the product of the devil."

Talk of the devil usually presses buttons in me, but Nazhar's gentle voice is reassuring, like the calm, silver-gray sea sliding by. Nazhar points to one of the islands. "They used cyanide over there, but now we are working together. They are asking us to train them in catching with the nets. With our limited materials and abilities, we try our best to teach them, teaching my people the value of the environment. We have been inculcating the good and bad of catching fish and other marine products."

I often wonder whether my conservation values are a luxury affordable only to middle-class white westerners. But these people do not have the securities I have, or the options, not are they being fed hysterical-sounding funding appeals by effete environmental groups. These are armed Muslims fighting for control of their own lives. In many ways, their interests are entirely different from mine, yet we have converged upon a similar assessment of the state of the reefs and the importance of the environment in securing human dignity and freedom. Interesting. So maybe conservation of nature is not just a Western, middle-class concept. Maybe it is a human need, a universal hunger.

Vaughan, still covered with mud and reminding me of how filthy we all have gotten today, is snoozing on a cot. How I envy him. Less than twenty-four hours ago I was in Hong Kong waiting for my canceled flight to be rescheduled. In the middle of the night I arrived in Manila with no sleep and got on another plane with Vaughan. Following today's events and last night's lack of sleep, I am being swept off my feet by a towering wave of exhaustion, a veritable tsunami of narcolepsy. Nazhar shows me to a bunk inside the wheelhouse, and I lay my wet, muddy, over-fatigued body down alongside two loaded Browning automatic rifles. My head comes to rest beside a copy of the *Holy Quer-An*. I am not in Kansas anymore.

Some time later, Nazhar is tapping my toe. Fourteen hours after leaving Manila—two airplanes, a truck, and a boat ride later—we have arrived at Dong-dong.

I force myself up and wobble outside. We are approaching a dock swelling with onlookers: children, women, men. Ropes are thrown and we tie up. Voices mingle. The air carries a hint of smoke. The kids—mostly grimy—are in shorts and T-shirts. A few kids under ten wear only their birthday suits. One girl holds a stringer of fish called rainbow runners. The shoreline—white sand, backed by coconut palms—is crowded with bamboo-and-thatch houses on stilts, extending out over clear water. Many other houses sit back from the beach in the shade of the palms. The smiling people who have come to see us ashore include a scattering of copiously munitioned riflemen.

New-looking, brightly painted outrigger boats rest upon the beach. These, I am told, were supplied by the cyanide purveyors. I am made to understand that the villagers have recently run those people out of town—and kept the boats.

Vaughan, several of the Ecology Foundation people, several guards, and I disembark, followed off the dock by a small crowd of onlookers. The village, with a couple of thousand inhabitants, presents immediate contrasts: houses over water, and houses on dry land; young men carrying arms, others playing volleyball; smiles of welcome, and squints of distrust. A young woman pauses and looks at me sweetly; another, slender and shapely with a toddler on her hip, sweeps by, glaring a frightening and inexplicable look of wild defiance.

The talk is in murmurs—except for an occasional shout on the volleyball court—but there is plenty of activity. Many people are busy spreading seaweed out to dry, and bales of it are stacked alongside some of the houses. Chickens strut in perpetual motion across thoroughfares of well-worn sand. People in some of the shacks are selling snacks and refreshments.

Our entourage is led to one house and invited to sit on the porch. The most influential man in the village, a holy man, praying upon his beads, sits next to us. He says nothing at all.

A basket of cigarettes is offered, "No, thank you," Prawn crackers? "Yes, thank you very much."

Some women, many of them quite young, nurse babies on the steps of the huts. Vaughan remarks, "Jeez, every girl who's old enough is pregnant." From 1966 to 1986, while the productivity of the coral reefs dropped by at least one-third, the population of the Philippines almost doubled. Because of these two trends, the square kilometer of reef that helped support nine hundred persons in 1966 had to help support twenty-six hundred persons in 1986, almost three times as many. The country's population is expected to double again in the first third of the twenty-first century.

Perfecto says to me, "This is the typical community that we are trying to work with. Our task is really very challenging. Can you imagine, reaching out to these deprived communities? It drains our organization's resources. But this is where the challenge is—to enhance skills, to preserve corals."

A circle of kids, young men, and women has moved quite close, staring at us intently and inspecting us minutely. Many of these kids have never seen outsiders. Vaughan says the kids are already asking the adults, "Why are these people here?" "Just by our arrival," he says, "they are learning about the corals, the fishes, the cyanide." In most places, Vaughan says, villagers don't fish near the village anymore. "The areas have been too mined. Those villages present too much of a logistical difficulty for us," he says. "Here, they still fish right near the village, so the entire spectrum of residents—elders, holy men, families, children—can be involved."

A couple of muscular young men bring a bunch of coconuts and set them down heavily, thudding the sand at their bare feet. Before Perfecto can say, "A welcoming drink," their machetes are whacking off the husks and opening the shell tops. Once we've received our coconuts, Perfecto smiles broadly, making a gesture of salute before we all take a deep sip. Vaughan says to me, loud enough for Perfecto to hear, "Without Pete to speak the local dialect and relate to the people, we wouldn't be able to work here."

A friendly man of about fifty sits down next to me as I am still sipping my coconut. He has an automatic rifle and a belt of cartridges. No big deal; no one gets shot by accident. Secondhand cigarette smoke seems to be a larger danger in this village.

Perfecto introduces us to this rifleman, explaining, "He is one of our local leaders. He is responsible for water and firewood." We smile and shake hands.

In the oncoming sunset, towering clouds are striking and piling up over the mountains in which the MNLF and the Abu Sayyaf rebels are hiding. The most heavily armed man I have ever seen, a cartridge belt loaded into an automatic rife and another one hung around his neck, accompanies us out to the dock. Around the village, cooking smokes are now coming up through the thatched roofs, scenting the breeze as families prepare evening meals. Filipino pop music comes filtering faintly from a battery-powered tape player in one of the houses.

By sundown about twenty of us are back aboard the boat. I say "us," but I have no idea who most of these people are. There are only two cots—which the hardworking women commandeer without contest. Most of the men sleep side by side on large mats. Deliriously sleepy but not quite ready to lie among strange men piled like puppies, I put on my sweat-shirt and rain gear, go up to the roof, lie on a towel, and shove a rolled-up T-shirt under my head. This expedition to Dong-dong, the grimy old wooden boat, the congressman's armies, cloud forests harboring revolutionaries and terrorists, the holy men, the disheveled thatch villages, chain-smoking men in fatigues dozing on mats surrounded by loaded rifles: It has seemed more an expedition into the heart of a world apart than an outing to—of all things—catch aquarium fish for living rooms in the United States. I would not have guessed that such seemingly frivolous hobbies could have at their origin such deadly serious business a world away. And so ends perhaps the longest—though for a while it seemed like it might become the shortest—and certainly the muddiest day of my life. I surrender my inert body to the dew drops.

First light: A slow-handed dawn brings the crowing of roosters and the Islamic call to the faithful. My hair and rain clothes are wet. I am chilly. But it takes the roaring sounds of the air compressors to bring me to full consciousness.

By the time I descend the ladder to the deck, we are already under way to the dive site. The engine-room mechanic, continually nursing the big diesel, controls the throttle based on signals from a makeshift bell that the captain rings by pulling a cord in the wheelhouse, making a clapper hit a suspended piece of pipe. The number of clangs directs the engine man to make the diesel run either faster or slower or slower still.

The air compressors are adding synergistically to the bone-vibrating noise of the engine. Scuba tanks are being carefully filled in a large plastic tub of cool water. These people have been trained well in scuba safety.

Perhaps a gun-safety course would be a good next project. The guards consistently walk with their guns horizontal, at every turn swiveling their line of fire through the midriffs of a dozen or more plain folk, including yours truly. Some like to keep a finger near the trigger; others like to rest the palm of a hand over the barrel. When a gun is placed at rest it is invariably horizontal. Men habitually lounge or doze with their own guns pointed into their brains, or at least their heads.

Some of the men are sorting dried, edible sea cucumbers. Our tireless young cook squats on deck, cleaning a dozen ballyhoo, her bare feet, as usual, covered with scales and slime and fish blood. Water for coffee is on the coals while a couple of fish bake alongside the kettle.

A few small flocks of sandpipers trickle by. I check their direction against the rising sun. North, among the most migratory of animals, these frailties are headed from this tropic theater to face spring snowstorms on Siberia's tundra, where they will breed.

Grilled fish and rice is served, but only to us, the outsiders. The fact that I prefer the fried, vinegar cassava with sour green mangoes and raw ballyhoo brings giggles of approval and thumbs-up from the locals, particularly the cook. They explain that this is their everyday breakfast food, not meant for guests, and they are very pleased that I'm enjoying it. "This is our bagel," explains one of the more worldly men to a New Yorker he perhaps thinks is Jewish.

The suspended pipe-bell in the engine room clangs several times, and the mechanic idles the Diesel's down. We are off the tip of an island. Nazhar comes over, pats my shoulder very gently, and says softly, "Here the coral reef is quite very O.K., not yet touched by the diving. We are looking for the future of our people right here."

Four outrigger canoes and one large outrigger skiff that have run here with us from Dong-dong gather around our big yellow boat. The boat, by the way, is nameless, for lack of consensus. I have suggested the names *Philippine Queen, Big Yellow Taxi,* and *The Yellow Submachine (Gun).* Inexplicably, neither Pratt nor Perfecto have embraced any of my lyrical suggestions.

The current is running very hard. Vaughan and I will use scuba gear, while the fish catchers will be breathing off the "hookahs": long air hoses connected to the compressor pump on their outrigger skiff. Two netters and I will go down together, and Vaughan will accompany another pair. The fish catchers are wearing sweatsuits and homemade plywood fins.

We enter the water and descend thirty feet at a tide-swept angle to a sandy flat interspersed with coral heads. These isolated heads are better than a large reef for capturing fish with nets because the mesh can be strung around the individual corals.

Fair numbers of fish are present, but—as Bob Johannes and Larry Sharron had warned about the Philippines—everything is under six inches. The fish catchers, trailing the hookah hoses from their mouths and breathing off the streaming air bubbles being pumped to them from the drifting outrigger skiff above, look like weird, hooked fish.

Even on the bottom, we must work hard just to avoid being swept away. I reach for a rock to hold myself in place, incurring an immediate sting on the wrist from a small coral. I decide to seek shelter from the current behind the bigger coral heads. Each time I move to a new head, I am confronted by a moray about a yard long, staring at me from the lee of the shelter. Green eels predominate, but one is mottled yellow and black: a snowflake moray. They've got the place staked out.

The current is playing havoc with the net, and the divers abandon a couple of initial attempts at a set. This dive could have been better timed to coincide with slack water, but the trainees, all former cyaniders, show remarkable persistence in getting the net to work right. For me, just holding in the current is difficult. If I were doing this for a living, and I had once used cyanide, the difficulty would be vexing. Working the net looks virtually impossible. But working they are, and working hard. They are endeavoring under conditions that are exactly wrong for encouraging early successes. These young men are heroes, showing an understanding, forbearance, and commitment to the future that I have found rare anywhere in the world—for example, New England and the Pacific Northwest in my own country, where people often behave as though they have stopped thinking about tomorrow.

One trainee works for ten minutes to entrap a single emperor angelfish. He finally succeeds, carefully removing the struggling blue booty from the net and slipping it into his mesh-topped bucket. This little emperor is worth about $12 to him. Someone in the U.S. will pay about $150 to acquire it.

Next, the net goes around a coral head of large size and complex structure. Its shadowed interstices harbor a potentially lucrative array of fishes. You can see them by peeking into the holes. But the fish understand their security, and despite the divers' attempts at frightening them out, they will not break cover. The other divers join the attempt to drive the fish out and into the net. They won't budge. After some minutes the frustrated fishers give up and move on.

We drift in the current, into an area lacking corals and seeming devoid of fish. The bottom here is pocked and careered. I look questioningly at Vaughan. He pantomimes an explanation, putting the knuckles of his fists together and then opening his palms and spreading his fingers while rotating his wrists outward. Explosion. Blast fishing. Much of the bottom is coral rubble, some of it still alive. But the structure, the coral community, is gone. I can see why many people have lost fingers, hands, and arms while they were doing this.

We find a standing coral head where the divers are lucky to get another emperor and one small sweetlips. The divers are very selective here, skipping numerous attractive fishes and shrimps and zeroing in on others. I wonder: Do they upset the ecology by removing these fish? A look around at the dynamite-blasted craters says to me, "Who cares if they do?" Compared to what has gone before, their determination with the nets is a vast improvement and a magnificent step forward to a simpler and better time.

One diver breaks a strap on his plywood fin and leaves for the surface. Another chases a beautiful yellow tang three times around a large coral head. The tang deftly avoids the net each time. The diver repositions the net and catches the fish, but it slips from his hand as he tries to put it in his zippered bucket, and it won't come out of cover again for anything.

He moves to a small, nondescript coral head that I would have swum right past. It turns out to be loaded: tiger-striped yellow pipefish, several kinds of shrimp, even a small grouper. While he is working, an anemonefish escapes from his bucket through an opening in the zipper. It is disoriented without its accustomed anemones; rather than darting away, it swims idly around us—*so* idly that I could catch it with my hand. The diver is so intent on catching rose-banded shrimps that he doesn't see the escaped fish hovering around him.

After sixty-six minutes underwater, we come to the surface. The take for the dive includes four emperor angels, several blue tangs, some domino damsels, anemonefishes, a bicolor angel, and three yellow tangs. This is the second net catch for these trainees. Experienced net catchers would get perhaps twice as many fish in a day (earning the equivalent of perhaps forty dollars each).

Ferdinand Cruz, International Marinelife Alliance's Cyanide Reform Program field director, has fought to get increased prices from Manila buyers for net-caught fish. Some have agreed. He says, "For blue tangs, fishermen used to get five pesos"—about twenty cents—"but now they get forty."

Vaughan says, "This way they obtain a comparably valued catch for fewer fish without cyanide, and what they like is that they can go back to the same coral head again and again, finding new fish there because the coral is alive and continues to draw and support them."

After a pause, Pratt continues, "The dealers who are making the big money selling these fish are doing none of the work. If people only knew what went into getting the fish they see and buy in aquarium shops."

Not least among the people who *are* doing the work, in my opinion, is Vaughan Pratt himself.

While many people talk about sustainable development, Pratt and his coworkers have proven that retraining, detection, and enforcement can work. These efforts may yet succeed in transforming the Philippines from the tropics' worst case of coastal destruction to a model of rehabilitation, and the work here in Sulu is not an isolated example. Pratt once took Bob Johannes and me on an expedition to Canipo Island in the Calamianes group in northern Palawan.

One fisherman there told me the cyanide-detection lab is the best thing that has happened. It is "a great deterrent," he said, "because the jail terms are so serious." And what is a serious jail term? Several people had recently gotten eight-year sentences. First offense.

Canipo Island was the most pleasant remote village I've ever been in, with real energy instead of the sense of stagnation that characterizes many. About sixty brightly painted and properly registered boats lay ready for launching along the white beach. Another, bigger, one was under construction. About as many houses, some with electricity, stood under the coconut palms. The thoroughfares—even the pigpens—were clean, and the woven bamboo houses, some of the them two stories high, showed much individual expression in size, style, and shape (one was even hexagonal). A bamboo hut had been erected for showing movies. Local fauna included cats, chickens, puppies, ducks, and a swarm of plump children.

The villagers had formed a co-op called Kamil Amianan, which translates to Hook and Line. I asked one of the local leaders, a man named Caesar, for the secret to the village's vitality.

"Before, when they were blast fishing with explosives made from fertilizer, our people could hardly afford paddleboats," he said. "Now, with Hook and Line, and with the live-fish market, we can afford to send our children to school." With the area protected from cyanide, the catches have stabilized, and the take seems sustainable now for both fish and people. "That's why this place is special."

Johannes was deeply impressed, offering, "I'm so glad I came here. In most places there is so much cyanide and destruction associated with the live trade, but at least now I can say I've been to a place where the people are fishing responsibly, looking to the future, and prospering. This is the most encouraging thing I've seen in a long time."

Caesar replied, "We fought to make it this way."

We dined at the home of the fishing co-op's head. The house had a swept-sand floor and an indoor toilet, and the fishermen generously lavished a delicious (and valuable) meal of red grouper upon us and housed us overnight on the island. The bedroom I was given featured pictures of nine Catholic saints above the headboard (some of whom I had not spoken with in years), two rifles under the bed, and a wall calendar decorated with twelve young women in poses that made me miss home.

Downstairs, one man was not being received as well. He was the captain of a large, dilapidated old fishing boat that, along with its impoverished crew of over eighty boys and men, was being detained by deputies overnight for fishing illegally. The captain told me he had come all the way from Mindoro, 120 miles north, because cyanide and blasting had destroyed their fishing grounds there. He would be released in the morning after promising not to return, but his worries seemed far from over.

The makeshift bell in the engine compartment clangs a couple of times as our own captain signals to the motorman to throttle the diesel down as we approach Dong-dong village. Again, about a hundred people gather on the dock to see the aquarium fishes we have captured today. One boatload of men, heavily armed, departs surreptitiously at our approach, but not surreptitiously enough that we don't notice them. It is the MNLF. They have been talking to Perfecto, who stayed onshore. They have heard we are here, and they want us to come to their village to train them to fish without cyanide.

On the dock, everyone is straining and jostling for a look into the buckets as the fish are sorted. The wildly improbable, gemlike colors and perpetual motion of the fresh-caught creatures captivate all, especially the children, putting them in a gleeful mood and giving them the giggles.

"This is environmental work at its best," Vaughan says. "It's really a social program."

The fish are transferred to perforated jars kept underneath the dock in the shade. A man points to one of the fish, saying to me proudly, "Thees ees thee clown treeggerfeesh." Happily, this catch is destined for the Ocean Park aquarium in Hong Kong, where these swimming jewels will continue to captivate people by the tens of thousands in the extraordinary new reef exhibit there (the largest in the world).

Pratt says, "The weakness for us is that the cyanide middlemen paid these catchers every day for the fish, and we can't do that yet."

Perfecto adds, "The issue here is survival, but the willingness for them to learn to fish clean is anchored on our ability to sustain a ready market for them. We need to set up co-ops and infrastructure from capture to market."

I ask Perfecto, who has been ashore all day, if he is coming back to *Big Yellow* with us. He seems to misunderstand the thrust of my question, and as several more heavily armed men board our boat, he says, "Yes, I am coming with you, but you did not have to worry, I had the Delta Force covering you all day."

We pull away from the dock and anchor a short distance from shore. New faces come and go continually in canoes. Onboard, amid the noise and smoke and smells of the deck, people are continually eating rice, cassava, fish—there is almost always something ready on the sooty fire. Our young cook works incessantly, with good humor and a quick-flash smile. She speaks no English, but after I try complimenting her cooking she makes a point of coming over to sit next to me in her rare moments off, gesturing requests for me to take more pictures of her with her boyfriend, who is part of the crew.

As the sun begins to settle, Nazhar himself settles next to me, saying softly, "You must come and visit us again in our beautiful place. Here, you have no worries. You are under our auspices, and no one will touch you. No one bad can get to you because you will always be under our protection."

Perfecto says to me, "The next time you visit, you must make more time to stay with us. I will show you more of our lovely corals. I will show you our wonderful mangroves. And—if you want—I will teach you how to shoot and disassemble an M16. It will be part of our cultural exchange." He chuckles.

By the time we finish eating and talking, this part of the world lies in darkness. I climb to the roof and look down into a galaxy of phosphorescence in the sea. In every heaven above float as many stars as seem possible to behold.

Three muffled shots report through the distant night. Someone paddling a canoe near the island is singing a strangely haunting melody in a voice full of beauty and all humanity, though the tongue to me is foreign. The song rolls across still waters and into the deep, blue-black universe beyond, a song from an ocean planet called Earth, a world of water, yet itself an island.

I spread my towel on the roof under the arching galaxy, lie down wearily in my clothes, and rest my eyes. I am thinking about irony, and hope, and the irony that hope is. The irony that ideals and hope can flourish at the tattered edges of human existence, the hope that if ideals can flourish here, they can melt civilization's jaded heart. Soon my thoughts begin to swim, and in a short while I am asleep.

The Party's Over

Oil, War, and the Fate of Industrial Societies

By Richard Heinberg

Introduction

In a previous chapter introduction, I suggested the water crisis would hit before global warming did. In *The Party's Over: Oil, War, and the Fate of Industrial Societies*, author Richard Heinberg suggests Peak Oil may already be upon us. This book, written in 2005, almost appears prophetic. There is too much year-to-year variation in oil production to be certain Peak Oil has occurred until several years after the fact; that said, the simultaneity of the oil and grain price jumps of 2008, coupled with the beginning of the Great Recession, is disturbing. Heinberg shows the ways in which fossil fuels (oil in particular) permeate every aspect of our lives; tracks the inevitable decline in oil production; demonstrates that the transition to renewable energy will not be easy; and predicts the devastating effects of Peak Oil on key sectors of human life, including the economy, transportation, agriculture, home heating and cooling, the environment, politics, and international resource competition.

The final sections I have chosen, from Heinberg's chapter "Managing the Collapse," describe the difficult but potentially rewarding responses to Peak Oil that are available to us at all levels, from personal to global. If *Our Global Future* has any main theme, it is richly expressed in this section of Heinberg's book. We face a future of limited means that will require greater attention to our local communities. The collapse of global society does not have to be the popping of a balloon; it can instead be the slow deflation of a dangerously overinflated economy. To use another analogy, imagine yourself walking across the room holding a very shallow, completely full tray of water. How would you avoid spilling it? Now imagine that same tray was an *ice cube* tray that you were walking to the refrigerator, full of water but with a separator breaking it down into separate sections (the "cubes"). That is an easy task by comparison. Our dangerously overheated global economy is similar to the first tray; money can freely slosh from one side to the other, with substantial wastage. Re-localization (installing the separator) would limit this movement, helping local economies.

Why this emphasis on localization? Globally, oil is perhaps 40% of our energy usage, but it comprises *90 percent of our transportation energy usage*. When the cost of oil goes through the roof, Chilean sea bass and Philippine mangoes will be a thing of the past here in the United States. Currently, food travels an average of 1500 miles farm-to-plate. This is wasteful and it will end, not because some environmental extremist (as those who advocate for sanity are so often characterized) forced government regulations upon an unwitting populace, but because it will become far too expensive. Heinberg suggests, as do I and many others, that this anti-globalization or localization, though painful, will in the long run prove far more sustainable, and may increase our sense of community. Alienation from our society, extreme boredom with our work, and oppression of our self-expression have been a big part of "civilized" life since the first irrigation civilizations evolved millennia ago; these factors have been amplified with the Industrial Revolution, as Congolese mine slaves and poisoned Mexican *maquiladora* workers and Marshallese textile sweatshop workers have poured their lifeblood into the global economy, with scant reward for products shipped far away. None of us can claim with any accuracy to have a window into the future, but it seems at least possible that a more local economy could be a more just one as well.

Lights Out

The Ground Giving Way

In nearly every year since 1859, the total amount of oil extracted from the world's ancient and finite underground reserves had grown—from a few thousand barrels a year to 65 million barrels per day by the end of the 20th century, an increase averaging about two percent per annum. Demand had grown just as dramatically, sometimes lagging behind the erratically expanding supply. The great oil crises of the 1970s—the most significant occasions when demand exceeded supply—had been politically based interruptions in the delivery of crude that was otherwise readily available; there had been no actual physical shortage of the substance then, or at any other time.

In the latter part of the year 2000, as Al Gore and George W. Bush were crisscrossing the nation vying for votes and campaign contributions, the world price of oil rose dramatically from its low point of $10 per barrel in February 1999 to $35 per barrel by mid-September of 2000. Essentially, Venezuela and Mexico had convinced the other members of OPEC to cease cheating on production quotas, and this resulted in a partial closing of the global petroleum spigot. Yet while Saudi Arabia, Iraq, and Russia still had excess production capacity that could have been brought on line to keep prices down, most other oil-producing nations were pumping at, or nearly at, full capacity throughout this period.

Meanwhile, a wave of mergers had swept the industry. Exxon and Mobil had combined into Exxon-Mobil, the world's largest oil company; Chevron had merged with Texaco; Conoco had merged with Phillips; and BP had purchased Amoco-Arco. Small and medium-sized companies—such as Tosco, Valero, and Ultramar Diamond Shamrock Corp—also joined in the mania for mergers, buyouts, and downsizing. Nationally, oil-company mergers, acquisitions, and divestments totaled $82 billion in 1998 and over $50 billion in 1999.

Altogether, the oil industry appeared to be in a mode of consolidation, not one of expansion. As Goldman Sachs put it in an August 1999 report, "The oil companies are not going to keep rigs employed to drill dry holes. They know it but are unable … to admit it. The great merger mania is nothing more than a scaling down of a dying industry in recognition that 90 percent of global conventional oil has already been found."

Meanwhile the Energy Information Agency (EIA) predicted that global *demand* for oil would continue to grow, increasing 60 percent by the year 2020 to roughly 40 billion barrels per year, or nearly 120 million barrels per day.

The dramatic price hikes of 2000 soon triggered a global economic recession. The link between energy prices and the economy was intuitively obvious and had been amply demonstrated by the oil crises and accompanying recessions of the 1970s. Yet, as late as mid-2000, many pundits were insisting that the new "information economy" of the 1990s was impervious to energy-price shocks. This trend of thought was typified in a comment by British Prime Minister Tony Blair, who in January 2000 stated that "[t]wenty years on from the oil shock of the '70s, most economists would agree that oil is no longer the most important commodity in the world economy. Now, that commodity is information."[6] Yet when fuel prices soared in Britain during the last quarter of the year, truckers went on strike, bringing commerce within that nation to a virtual standstill. Though energy resources now *directly* accounted for only a small portion of economic activity in industrialized countries—1.2 percent to 2 percent in the US—all manufacturing and transportation still required fuel. In fact, the *entire* economy in every industrial nation was completely dependent on the continuing availability of energy resources at low and stable prices.

As the world economy slowed, demand for new goods also slowed, and manufacturing and transportation were scaled back. As a result, demand for oil also decreased, falling roughly five percent in the ensuing year. Prices for crude began to soften.

Indeed, by late 2001, oil prices had plummeted partly as the result of market-share competition between Russia and Saudi Arabia. Gasoline prices at the pump in California had topped $2 in late 2000, but by early 2002 they had drifted to a mere $1.12 per gallon.

Such low prices tended to breed complacency. The Bush administration warned of future energy shortages, but proposed to solve the problem by promoting exploration and production within the US and by building more nuclear power plants—ideas that few with much knowledge of the energy industry took seriously. Now that gasoline prices were low again, not many citizens contemplated the possible future implications of the price run-ups of 2000. In contrast, industry insiders expressed growing concern that fundamental limits to oil production were within sight.

This concern gained public recognition in 2004, as oil prices again shot upward, this time attaining all-time highs of over $55 per barrel. *National Geographic* proclaimed in its cover story that this was "The End of Cheap Oil"; *Le Monde* announced "The Petro-Apocalypse;" while Paul Erdman, writing for the CBS television magazine *Marketwatch*, proclaimed that "the looming oil crisis will dwarf 1973." In article after article, analysts pointed to dwindling discoveries of new oil, evaporating spare production capacity, and burgeoning global demand for crude. The upshot: world oil production might be near its all-time peak.

If this were indeed the case—that world petroleum production would soon no longer be able to keep up with demand—it would be the most important news item of the dawning century, dwarfing even the atrocities of September 11. Oil was what had made 20th-century industrialism possible; it was the crucial material that had given the US its economic and technological edge during the first two-thirds of the century, enabling it to become the world's superpower. If world production of oil

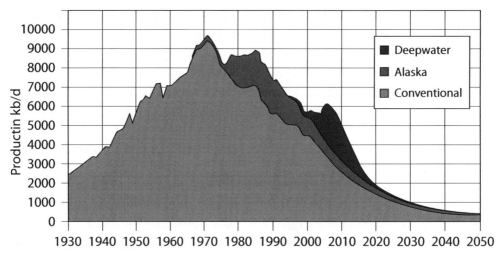

Figure 5. US oil production, history and projection, including lower 48, Alaska and Gulf of Mexico (deep water). (Source: Campbell and Laherrere, Association for the Study of Peak Oil & Gas.)

could no longer expand, the global economy would be structurally imperiled. The implications were staggering.

There was every reason to assume that the Bush administration understood at least the essential outlines of the situation. Not only were many policy makers themselves—including the President, Vice President, and National Security Advisor—former oil industry executives; in addition, Vice President Dick Cheney's chief petroleum-futures guru, Matthew Simmons, had warned his clients of coming energy-supply crises repeatedly. Moreover, for many years the CIA had been monitoring global petroleum supplies; it had, for example, subscribed to the yearly report of Switzerland-based Petroconsultants, published at $35,000 per copy, and was thus surely also aware of another report, also supplied by Petroconsultants, titled "The World's Oil Supply 1995," which predicted that the peak of global oil production would occur during the first decade of the new century.

It would be an understatement to say that the general public was poorly prepared to understand this information or to appreciate its gravity. The *New*

York Times had carried the stories of the oil company mergers on its front pages, but offered its readers little analysis of the state of the industry or that of the geological resources on which it depended. Mass-audience magazines *Discover* and *Popular Science* blandly noted, in buried paragraphs or sidebars, that "early in [the new century] ... half the world's known oil supply will have been used, and oil production will slide into permanent decline" and that "experts predict that production will peak in 2010, and then drop over subsequent years"—but these publications made no attempt to inform readers of the monumental implications of these statements. It would be safe to say that the average person had no clue whatever that the entire world was poised on the brink of an economic cataclysm that was as vast and unprecedented as it was inevitable.

Yet here and there were individuals who did perfectly comprehend the situation. Many were petroleum geologists who had spent their careers searching the globe for oil deposits, honing the theoretical and technical skills that enabled them to assess fairly accurately just how much oil was left in the ground, where it was located, and how easily it could be accessed.

What these people knew about the coming production peak—and how and when they arrived at this knowledge—constitutes a story that centers on the work of one extraordinary scientist.

M. King Hubbert: Energy Visionary

During the 1950s, '60s, and '70s, Marion King Hubbert became one of the best-known geophysicists in the world because of his disturbing prediction, first announced in 1949, that the fossil-fuel era would prove to be very brief.

Of course, the idea that oil would run out eventually was not, in itself, original. Indeed, in the 1920s many geologists had warned that world petroleum supplies would be exhausted in a matter of years. After all, the early wells in Pennsylvania had played out quickly; and extrapolating that initial experience to the limited reserves known in the first two decades of the century yielded an extremely pessimistic forecast for oil's future. However, the huge discoveries of the 1930s in east Texas and the Persian Gulf made such predictions laughable. Each year far more oil was being found than was being extracted. The doomsayers having been proven wrong, most people associated with the industry came to assume that supply and demand could continue to increase far into the future, with no end in sight. Hubbert, armed with better data and methods, doggedly challenged that assumption.

M. King Hubbert had been born in 1903 in central Texas, the hub of world oil exploration during the early 20th century. After showing a childhood fascination with steam engines and telephones, he settled on a career in science. He earned BS, MS, and Ph.D. degrees at the University of Chicago and, during the 1930s, taught geophysics at Columbia University. In the summer months, he worked for the Amerada Petroleum Corporation in Oklahoma, the Illinois State Geological Survey, and the United States Geological Survey (USGS). In 1943, after serving as a senior analyst at the Board of Economic Warfare in Washington, DC, Hubbert joined Shell Oil Company in Houston, where he directed the Shell research laboratory. He retired from Shell in 1964, then joined the USGS as a senior research geophysicist, a position he held until 1976. In his later years, he also taught occasionally at Stanford University, the University of California at Los Angeles, the University of California at Berkeley, the Massachusetts Institute of Technology, and Johns Hopkins University.

During his career, Hubbert made many important contributions to geophysics. In 1937 he resolved a standing paradox regarding the apparent strength of rocks that form the Earth's crust. Despite their evident properties of hardness and brittleness, such rocks often show signs of plastic flow. Hubbert demonstrated mathematically that, because even the hardest of rocks are subject to immense pressures at depth, they can respond in a manner similar to soft muds or clays. In the early 1950s, he showed that underground fluids can become entrapped under circumstances previously not thought possible, a finding that resulted in the redesign of techniques employed to locate oil and natural gas deposits. And by 1959, in collaboration with USGS geologist William W. Rubey, Hubbert also explained some puzzling characteristics of overthrust faults—low-angle fractures in rock formations in which one surface is displaced relative to another by a distance on the order of kilometers.

These scientific achievements would have been sufficient to assure Hubbert a prominent place in the history of geology. However, Hubbert's greatest recognition came from his studies of petroleum and natural gas reserves—studies he had begun in 1926 while a student at the University of Chicago. In 1949, he used statistical and physical methods to calculate total world oil and natural gas supplies and documented their sharply increasing consumption.

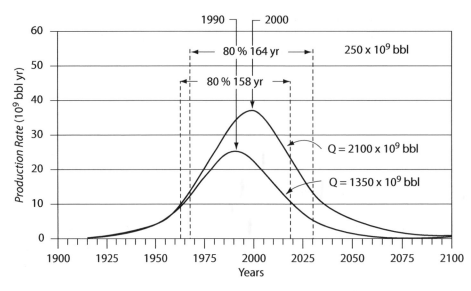

Figure 6. M. King Hubbert's projected cycles for world crude production for the extreme values of the estimated total resource. (Source: M. K. Hubbert, Resources and Man)

Then, in 1956, on the basis of his reserve estimates and his study of the lifetime production profile of typical oil reservoirs, he predicted that the peak of crude-oil production in the United States would occur between 1966 and 1972. At the time, most economists, oil companies, and government agencies (including the USGS) dismissed the prediction. The actual peak of US oil production occurred in 1970, though this was not apparent until 1971.

Let us trace just how Hubbert arrived at his prediction. First, he noted that production from a typical reservoir or province does not begin, increase to some stable level, continue at that level for a long period, and then suddenly drop off to nothing after all of the oil is gone. Rather, production tends to follow a bell-shaped curve. The first exploratory well that punctures a reservoir is capable of extracting only a limited amount; but once the reservoir has been mapped, more wells can be drilled.

During this early phase, production increases rapidly as the easiest-accessed oil is drained first. However, beyond a certain point, whatever remains is harder to get at. Production begins to decline,

even if more wells are still being drilled. Typically, the production peak will occur when about half of the total oil in the reservoir has been extracted. Even after production has tapered off, some oil will still be left in the ground: it is economically impractical—and physically impossible—to remove every last drop. Indeed, for some reservoirs only a few percent of the existing oil may be recoverable (the average is between 30 and 50 percent).

Hubbert also examined the history of discovery in the lower-48 United States. More oil had been found in the 1930s than in any decade before or since—and this despite the fact that investment in exploration had increased dramatically in succeeding decades. Thus discovery also appeared to follow a bell-shaped curve. Once the history of discovery had been charted, Hubbert was able to estimate the total ultimately recoverable reserves (URR) for the entire lower-48 region. He arrived at two figures: the most pessimistic reasonable amount (150 billion barrels) and the most optimistic reasonable amount (200 billion barrels). Using these two estimates, he calculated future production rates. If the total

URR in the lower-48 US amounted to 150 billion barrels, half would be gone—and production would peak—in 1966; if the figure were closer to 200 billion barrels, the peak would come in 1972.

These early calculations involved a certain amount of guesswork. For example, Hubbert chose to chart production rates on a logistic curve, whereas he might have employed a better-fitting Gaussian curve. Even today, according to Princeton University geophysicist Kenneth S. Deffeyes, author of *Hubbert's Peak: The Impending World Oil Shortage,* the "numerical methods that Hubbert used to make his prediction are not crystal clear." Despite many conversations with Hubbert and ensuing years spent attempting to reconstruct those original calculations, Deffeyes finds aspects of Hubbert's process obscure and "messy." Nevertheless, Hubbert did succeed in obtaining important, useful findings.

Following his prediction of the US production peak, Hubbert devoted his efforts to forecasting the global production peak. With the figures then available for the likely total recoverable world petroleum reserves, he estimated that the peak would come between the years 1990 and 2000. This forecast would prove too pessimistic, partly because of inadequate data and party because of minor flaws in Hubbert's method. Nevertheless, as we will see shortly, other researchers would later refine both input data and method in order to arrive at more reliable predictions—ones that would vary only about a decade from Hubbert's.

Hubbert immediately grasped the vast economic and social implications of this information. He understood the role of fossil fuels in the creation of the modern industrial world, and thus foresaw the wrenching transition that would likely occur following the peak in global extraction rates. In lectures and articles, starting in the 1950s, Hubbert outlined how society needed to change in order to prepare for a post-petroleum regime. The following passage, part of a summary by Hubbert of one of his own lectures, conveys some of the breadth and flavor of his macrosocial thinking:

The world's present industrial civilization is handicapped by the coexistence of two universal, overlapping, and incompatible intellectual systems: the accumulated knowledge of the last four centuries of the properties and interrelationships of matter and energy; and the associated monetary culture which has evolved from folkways of prehistoric origin.

The first of these two systems has been responsible for the spectacular rise, principally during the last two centuries, of the present industrial system and is essential for its continuance. The second, an inheritance from the prescientific past, operates by rules of its own having little in common with those of the matter-energy system. Nevertheless, the monetary system, by means of a loose coupling, exercises a general control over the matter-energy system upon which it is superimposed.

Despite their inherent incompatibilities, these two systems during the last two centuries have had one fundamental characteristic in common, namely exponential growth, which has made a reasonably stable coexistence possible. But, for various reasons, it is impossible for the matter-energy system to sustain exponential growth for more than a few tens of doublings, and this phase is by now almost over. The monetary system has no such constraints, and, according to one of its most fundamental rules, it must continue to grow by compound interest.

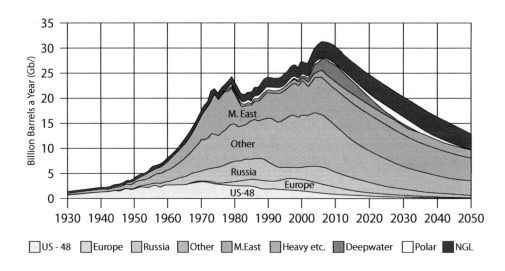

Figure 7a. World oil production, history and projection. (Source: Campbell and Laherrere, Association for the Study of Peak Oil & Gas.)

Amount		Gb	Annual Rate—Regular Oil					Gb	Peak
Regular Oil			Mb/d	2005	2010	2020	2050	Total	Date
Past	Future	Total	US-48	3.4	2.7	1.7	0.4	200	1972
Known Fields	New		Europe	5.2	3.6	1.8	0.3	75	2000
945	770	135	1850	Russia	9.1	8	5.4	1.5	210	1987
905			ME Gulf	20	20	20	12	675	1974
All Liquids			Other	29	25	17	8	690	2004
1040	1360	2400	World	66	60	46	22	1850	2006
2004 Base Scenario			Annual Rate—Other						
M. East producing at capacity (anomalous reporting corrected)			Heavy etc.	2.4	4	5	4	160	2021
			Deepwater	5.6	9	4	0	58	2009
Regular 0/7 excludes oil from coal, shale, bitumen, heavy deepwater, polar & gasfield NGL			Polar	0.9	1	2	0	52	2030
			Gas Liquid	8.0	9	10	8	275	2027
			Rounding		2	-2		5	
Revised 26/12/2004			ALL	83	85	65	35	2400	2007

Figure 7b. Estimated world oil production to 2100. (Source: Campbell and Laherrere, Association for the Study of Peak Oil & Gas.)

Hubbert thus believed that society, if it is to avoid chaos during the energy decline, must give up its antiquated, debt-and-interest-based monetary system and adopt a system of accounts based on matter-energy—an inherently ecological system that would acknowledge the finite nature of essential resources.

Hubbert was quoted as saying that we are in a "crisis in the evolution of human society. It's unique to both human and geologic history. It has never happened before and it can't possibly happen again. You can only use oil once. You can only use metals once. Soon all the oil is going to be burned and all the metals mined and scattered."

Statements like this one gave Hubbert the popular image of a doomsayer. Yet he was not a pessimist; indeed, on occasion he could assume the role of Utopian seer. We have, he believed, the necessary know-how; all we need do is overhaul our culture and find an alternative to money. If society were to develop solar-energy technologies, reduce its population and its demands on resources, and develop a steady-state economy to replace the present one based on unending growth, our species' future could be rosy indeed. "We are not starting from zero," he emphasized. "We have an enormous amount of existing technical knowledge. It's just a matter of putting it all together. We still have great flexibility but our maneuverability will diminish with time."

Reading Hubbert's few published works—for example, his statement before the House of Representatives Subcommittee on the Environment on June 6, 1974—one is struck by his ability to follow the implications of his findings on oil depletion through the domains of economics and ecology.[15] He was a holistic and interdisciplinary thinker who deserves, if anyone does, to be called a prophet of the coming era.

Hubbert died in 1989, a few years before his predicted date for the global production peak. That all-important forecast date was incorrect, as the rate of world oil production continued to increase through the first months of 2005. But by how far did he miss the mark? It would be up to his followers to find out.

Hubbert's Legacy

Since Hubbert's death, several other prominent petroleum geologists have used their own versions of his method to make updated predictions of the world's oil production peak. Their results diverge only narrowly from one another's. Since these scientists have been able to maintain updated data on reserves and production rates and since their work figures prominently in the current discussion about petroleum depletion, it will be helpful to introduce some of these individuals.

Colin J. Campbell is by most accounts the dean among Hubbert's followers. After earning his Ph.D. at Oxford in 1957, Campbell worked first for Texaco and then Amoco as an exploration geologist, his career taking him to Borneo, Trinidad, Colombia, Australia, Papua New Guinea, the US, Ecuador, the United Kingdom, Ireland, and Norway. He later was associated with Petroconsultants in Geneva, Switzerland, and in 2001 brought about the creation of the Association for the Study of Peak Oil (ASPO), which has members affiliated with universities in Europe. He has published extensively on the subject of petroleum depletion, and is the author of the book *The Coming Oil Crisis.*

Campbell's most prominent and influential publication was the article, "The End of Cheap Oil?", which appeared in the March 1998 issue of *Scientific American.* The co-author of that article, Jean Laherrère, has worked for the oil company Total (now Total Fina Elf) for thirty-seven years in a variety of roles encompassing exploration activities in the Sahara, Australia, Canada, and Paris. Like Campbell, Laherrère had also been associated with Petroconsultants in Geneva.

The *Scientific American* article's most arresting features were its sobering title and its conclusion:

> From an economic perspective, when the world runs completely out of oil is … not directly relevant: what matters is when production begins to taper off. Beyond that point, prices will rise unless demand declines commensurately. Using several different techniques to estimate the current reserves of conventional oil and the amount still left to be discovered, we conclude that the decline will begin before 2010.

Can the Party Continue?

With such a broad array of alternatives to choose from, many people assume it must be possible to cobble together a complex strategy to enable a relatively painless transition away from fossil fuels. Surely, for example, by building more wind turbines and fuel cells, by exploiting advances in photovoltaic technologies, and by redoubling our national conservation efforts, we could effortlessly weather the downside of the Hubbert curve.

A recurring subtext of this chapter has been the importance of net-energy analysis. To date, very few such analyses have been performed by impartial and competent parties. It is essential to the welfare of current and future generations that a standardized and well-defined net-energy methodology be adopted by national and international planning agencies. Reliance on market price as a basis for energy policy is shortsighted, because hidden subsidies so often distort the picture. Any standardized EROEI [Energy Returned on Energy Invested] evaluation methodology will inevitably be imperfect, but it will nevertheless provide the public and decision makers alike with much sounder insights into the costs of various energy options before precious resources are committed to them. As we have seen, the net-energy returns for some renewables (particularly wind) already exceed the dwindling returns for nonrenewable coal and domestic petroleum. Other options (such as hydrogen) may lose their luster when looked at closely.

Clearly, we would see the best outcome if all of the nations of the world were to undertake a full-scale effort toward conservation and the transition to renewables, beginning immediately. And undoubtedly some sort of complex strategy will eventually be adopted. But we should not delude ourselves. Any strategy of transition will be costly—in terms of dollars, energy, and/or our standard of living. Odum and Odum summarize the situation succinctly: "Although many energy substitutions and conservation measures are possible, none in sight now have the quantity and quality to substitute for the rich fossil fuels to support the high levels of structure and process of our current civilization."[44]

This is somewhat of a double message. Renewable alternatives are capable of providing net-energy benefit to industrial societies. We *should* be investing in them and converting our infrastructure to use them. If there is any solution to industrial societies' approaching energy crises, renewables plus conservation will provide it. Yet in order to achieve a transition from nonrenewables to renewables, decades will be required—and we do not have decades before the peaks in the extraction rates of oil and natural gas occur. Moreover, even in the best case, the transition will require shifting investment from other sectors of the economy (such as the military) toward energy research, conservation, and the implementation of renewable alternatives. Those alternatives will be unable to support the *kinds* of transportation, food, and dwelling infrastructure we now have; thus the transition will necessarily be comprehensive: it will entail an almost complete redesign of industrial societies. The result—an energy-conserving society that is less mobile, more localized, and more materially modest—may bring highly desirable lifestyle benefits for our descendants. Yet it is misleading to think that we can achieve that result easily or painlessly.

A Banquet of Consequences

Clearly, the energy transition of the early 21st century will affect nearly everything that humans care about. No person or group will be untouched by this great watershed.

Because the shift will be incremental, it would be a mistake to assume that the effects discussed in this chapter will all occur soon or in an instantaneous fashion. However, it would also be a mistake to assume that they will be so gradual in their appearance that they will accumulate to truly dramatic proportions only in our grandchildren's lifetimes or later. The early effects of the net-energy peak are already upon us and will probably begin to cascade within the next two decades or even the next few years.

Industrial civilization is a complexly interrelated entity, and it will respond to the net-energy decline as a system. It is therefore problematic to deal separately with effects on agriculture, transportation, and economics because developments in any one area will impact—and be impacted by—developments elsewhere in the system. However, written information must necessarily be organized in a linear fashion, so we will deal with various aspects of society one by one. As we do so, the reader may wish to give some thought to the ways in which each aspect dovetails with the others.

The Economy—Physical and Financial

The links between the physical economies of nations—their production, distribution, and consumption of goods and services—and the availability of energy are fairly obvious, but the subject bears some discussion nevertheless. All human activities require energy, which physicists define as the capacity to do work. With less net energy available, less work can be done—unless the efficiency of the process of converting energy to work is raised at the same rate as that at which energy availability declines. It will therefore be essential, over the next few decades, for all economic processes to be made as energy-efficient as possible. However, as discussed in the previous chapter, efforts to improve efficiency are subject to diminishing returns, and so eventually a point will be reached when reduced energy availability will translate into reduced economic activity.

Our current financial system was designed during a period of consistent growth in available energy, with its designers operating under the assumption that continued economic growth was both inevitable and desirable. This *ideology* of growth has become embodied in systemic financial structures *requiring* growth. The most prominent of these is compound interest.

Suppose you were to deposit $100 in a bank account earning six percent interest, and left it there for your children or grandchildren. After the first year, you would have $106, and after the second, $112.30. In twelve years your deposit would have doubled, and in a hundred years it would grow to $33,930. Unfortunately, the compound interest on debt works the same way: if you were to take out a loan of $100 at six percent interest and fail to make payments, over time that debt would grow similarly.

Currently all nations have a type of monetary system in which virtually all money is created through the making of loans. Thus, nearly all of the money in existence represents debt. For those not familiar with banking, this may be a difficult fact to grasp: I find that when I present it to college students, I often have to reiterate it in various ways for an hour or so before they are able to comprehend that money is not a physical substance kept in a vault, but a fictitious entity created out of nothing by bankers in order to facilitate the keeping of accounts.

All of this being so, a problem arises: From where does the money come with which to pay back the *interest* on loans? Ultimately, that money has to come from new loans, taken out by others somewhere else within the financial network of the economy. If new loans are not being made, then somewhere in the network people will be finding it impossible to pay the interest on their existing loans, and bankruptcies will follow. Thus the necessity for growth in the money supply is a structural feature of the financial system. The system seems to function best when growth in the money supply is kept at a low and fairly constant rate, and this is the job of the national banks (in the US, the Federal Reserve; in Canada, the Bank of Canada; in England, the Bank of England, and so on), which adjust interest rates to this end.

If money creation (i.e., the making of loans) occurs more rapidly than the growth in the production and consumption of goods and services in the economy, then *inflation* results; money then has less purchasing power, and this is bad for lenders—since the money used to repay loans is then worth less than the money that was borrowed. If money is not being loaned out (i.e., created) at a pace fast enough to match the growth in goods and services, then not enough money will be available to repay existing loans (plus interest), and the resulting bankruptcies and foreclosures can—in extreme cases—cause the economy to go into a tailspin of cascading financial cannibalism. *Deflation* may ensue, in which the purchasing power of money actually increases.

Until now, this loose linkage between a financial system predicated upon the perpetual growth of the money supply and an economy growing year by year because of an increasing availability of energy and other resources has worked reasonably well—with a few notable exceptions, such as the Great Depression. Productivity—the output produced per worker-hour—has grown dramatically, not because workers have worked harder but because workers have been controlling ever more energy in order to accomplish their tasks. Productivity, total economic activity, population, and money supply have all grown—at rates that have fluctuated, but within acceptable ranges.

The lower-energy economy of the future will be characterized by lowered productivity. There could be a good side to this in that more human labor will be required in order to do the same amount of work, with human muscle-power partially replacing the power of fossil fuels. Theoretically, this could translate into near-zero unemployment rates.

However, the financial system may not respond rationally. With less physical economic activity occurring, businesses would be motivated to take out fewer loans. This might predictably trigger a financial crisis, which would in turn likely undermine any attempts at a smooth economic adjustment.

As Hubbert pointed out, the linkage between the money system (the financial economy) and the human matter-energy system (the physical economy) is imperfect. It is possible for a crisis to occur in the financial system even when energy, raw materials, and labor remain abundant, as happened in the 1930s. But is it also possible for the financial system to remain healthy through an energy-led decline in the physical economy? That, unfortunately, is highly unlikely, due to the dependence of the former on continued borrowing to finance activity in the latter. Rather, it is highly likely that the net-energy decline will sooner or later trigger a financial crisis through a reduction in demand for goods and services, and hence for money (via loans) with which to pay for the machinery to produce those goods and services. Thus even if human labor is sufficiently abundant to make up for some of the reduction work performed by fuel-burning machines, the financial system may not be able to adapt quickly enough to provide employment for potential laborers.

Therefore extreme dislocations in both the financial system and the human matter-energy economy are likely during the energy transition. The exact form these dislocations will take is difficult to foresee. Efforts could be made to artificially pump up the financial system through government borrowing—perhaps to finance military adventures. Such massive, inflationary borrowing might flood markets with money that would be losing its value so quickly as to become nearly worthless. On the other hand, if inflationary efforts are not undertaken quickly or strenuously enough when needed, then the flagging rate of loans might cause money to disappear from the economy; in that case, catastrophic deflation would result. As was true in the Great Depression, what little money was available would have high purchasing power, but there would simply be too little of it to go around. Unemployment, resource and product shortages, bankruptcies, bank failures, and mortgage foreclosures would proliferate.

It is entirely possible that, over a period of decades, both inflationary and deflationary episodes may occur; however, due to the lack of a stable linkage between money and energy, periods of financial stability will likely be rare and brief.

Continued population growth, even at reduced rates, will put added strain on support systems and exacerbate the existing inherent requirement for economic growth.

Who will feel the pain? Most likely, the poor will feel it first and hardest. This will probably be true both nationally and internationally, as rich nations will likely seek to obtain energy resources from the poorer nations that have them by financial chicanery or outright military seizure. Eventually, however, everyone will be affected.

Some comforts, even luxuries, will probably continue to be available in most countries; but regardless of whether the financial environment is inflationary or deflationary, nearly everything that is genuinely useful will become relatively more expensive because the energy employed in its extraction or production will have grown more rare and valuable.

Transportation

The automobile is one of the most energy-intensive modes of transportation ever invented. This is true not just because of its direct use of fuel (a lightly loaded bus, airliner, or train actually uses more fuel per passenger-mile) but for the energy embodied in the construction of so many individual units that require replacement every few years. The rate of car ownership in the US is now 775 per thousand people—nearly the highest in the world—and many less-consuming nations, such as China, are foolishly seeking to emulate the American love affair with the automobile. Because increased car ownership results in changed patterns of urban development and resource distribution, it creates social dependency. Wherever this dependency has taken hold, it will have ruinous consequences in the coming century.

Over the short term, more energy-efficient cars will be built, including gasoline-electric hybrids, and perhaps some hydrogen-powered models. But the relentless economics of the energy decline will mean that—eventually but inevitably—fewer cars will be built. Only the wealthy will be able to afford them. The global fleet of autos will gradually age and diminish in number through attrition. For a peek at the year 2050, look to Cuba—where 50-year-old Fords and Plymouths are still in service because virtually no newer ones have been imported from the US due to the trade embargo.

During the 20th century, millions of miles of roads and highways were built for automobile and truck traffic, at extraordinary expense. The Los Angeles Freeway, for example, cost taxpayers $127 million per mile to construct. In fiscal 1995 alone, local, state, and federal governments in the US spent $8 billion on roads and highways. Last century's prodigious road-building feat—dwarfing any of the wonders of the ancient world—was only possible because oil was cheap. Asphalt incorporates large quantities of oil, and road-building machines run on refined petroleum. In the decades ahead, road building will grind to a halt and existing roads will gradually disintegrate as even repair efforts become unaffordable.

Countries with good public transportation—street cars, buses, subway and trains—will be much better poised than the US to weather the energy transition. Mass-transit users typically spend $200 to $2,000 per year for travel, considerably less than car owners spend. Also, when well utilized, mass-transit systems consume much less energy per passenger mile than do automobiles. In her book *Divorce Your Car! Ending the Love Affair with the Automobile*, Katie Alvord points out that "[w]hile a single automobile uses over 5,000 BTUs per passenger mile, a train car carrying 19 people uses about 2,300 and a bus carrying the same number only about 1,000."[1] However, the construction of mass-transit systems itself requires a sizable energy investment.

The tourism industry will languish in the decades ahead. This could have devastating effects on places like Hawaii, whose economies are almost entirely dependent on tourism.

But even more serious consequences of reduced transportation will be felt in disruptions in the distribution of goods. In the 1980s and '90s, increased global trade resulted in the moving of products and raw materials ever further distances from source to end user. As transportation fuels dwindle—for air, sea, and land travel—we will see an inevitable return to local production for local consumption. But this process of "globalization in reverse" will not be without difficulty, since local production infrastructures were often cannibalized in the building

of the global economy. For example, no large shoe companies continue to manufacture their products in the US. Unfortunately, the rebuilding of local production infrastructures will be problematic with less energy available.

Food and Agriculture

Throughout the 20th century, food production expanded dramatically in country after country, and virtually all of this increase was directly or indirectly attributable to energy inputs. Since 1940, the productivity of US farmland has grown at an average rate of two percent per year—roughly the same pace as that by which oil consumption has increased. Overall, global food production approximately tripled during the 20th century, just keeping pace with population growth.

Modern industrial agriculture has become energy-intensive in every respect. Tractors and other farm machinery burn diesel fuel or gasoline; nitrogen fertilizers are produced from natural gas; pesticides and herbicides are synthesized from oil; seeds, chemicals, and crops are transported long distances by truck; and foods are often cooked with natural gas and packaged in oil-derived plastics before reaching the consumer. If food-production efficiency is measured by the ratio between the amount of energy input required to produce a given amount of food and the energy contained in that food, then industrial agriculture is by far the least efficient form of food production ever practiced. Traditional forms of agriculture produced a small solar-energy surplus: each pound of food contained somewhat more stored energy from sunlight than humans, often with the help of animals, had to expend in growing it. That meager margin was what sustained life. Today, from farm to plate, depending on the degree to which it has been processed, a typical food item may embody input energy between four and several hundred times its food energy. This energy deficit can only be maintained because of the availability of cheap fossil fuels, a temporary gift from the Earth's geologic past.

While the application of fossil energy to farming has raised productivity, income to farmers has not kept pace. For consumers, food is cheap; but farmers often find themselves spending more to produce a crop than they can sell it for. As a result, many farmers have given up their way of life and sought urban employment. In industrialized countries, the proportion of the population that farms full-time fell precipitously during the 20th century. In 1880, 70.5 percent of the population of the United States were rural; by 1910, the rural population had already declined to 53.7 percent. In the US today, there are so few full-time farmers that census forms for the year 2000 included no such category in their list of occupations.

Mechanization favors large-scale farming operations. In 1900, the average size of a farm in Iowa was 150 acres; in 2000 it was well over twice that figure. However, the proportion of food produced by family farmers on a few hundred acres is itself dwindling; the trend is toward production by agribusiness corporations that farm thousands, even tens or hundreds of thousands of acres. In addition, a few giant multinational corporations control the production and distribution of seed, agricultural chemicals, and farm equipment, while other huge corporations control national and international crop wholesaling.

The transportation of food ever further distances has led to the globalization of food systems. Rich industrialized nations have used loans, bribes, and military force to persuade nations with indigenous populations surviving on small-scale, traditional subsistence cultivation to remove peasants from the land and grow monocrops for export. In the early part of the 20th century this practice gave rise to the phrase "banana republic," but the latter half of the century only saw the trend increase. In nation after nation, tiny subsistence plots were joined together into huge corporate-owned plantations producing coffee, tea, sugar, nuts, or tropical fruits for consumers in the US, Europe, and the increasingly prosperous countries of the Far East. Meanwhile, the ranks of the urban poor grew as peasants from the countryside flocked to shantytowns on the outskirts of places like Mexico City, Lagos, Sao Paulo, and Djakarta.

Today in North America, food travels an average of 1,300 miles from farm to plate. Consumers in Minneapolis and Toronto enjoy mangoes, papayas, and avocados year-round. In London, butter from New Zealand is cheaper than butter from Devon.

The production of meat and the harvesting of fish have likewise resulted in more energy consumption over the course of recent decades. A carnivorous diet is inherently more energy-intensive than a vegetarian diet; as growing populations in the Americas and Asia have adopted a more meat-centered fast-food diet, energy inputs per average food calorie have increased. Motorized fishing boats are much more effective at harvesting fish from the sea than their 19th-century sailing equivalents, though they are far less energy-efficient. But their very effectiveness poses a problem in that nearly all marine fisheries are now in decline as a result of overfishing.

The ecological effects of fossil fuel–based food production have been catastrophic, particularly with respect to agriculture. Farmers now tend to treat soil as an inert medium with which to prop up plants while force-feeding them chemical nutrients. As a result, the complex ecology of the living soil is being destroyed, leading to increased wind and water erosion. For every bushel of corn produced in Iowa, three bushels of topsoil are lost forever. Meanwhile, agricultural chemicals pollute lakes, rivers, and streams, contributing to soaring extinction rates among mammals, birds, fish, and amphibians.

There are signs that limits to productivity increases from industrial agriculture are already well within sight. Global per-capita food production has been falling for the past several years. Grain surpluses in the exporting countries (Canada, the US, Argentina, and the European Union) relative to global demand have disappeared, and farmers are finding it increasingly difficult to maintain production rates of a range of crops due to the salinization of irrigated croplands, erosion, the loss of pollinator species, evolved chemical resistance among pests, and global warming. For each of the past several years, world grain production has failed to meet demand, and grain in storage is being drawn down at a rate such that stocks will be completely depleted within two to five years.

Prospects for increasing food production above the global level of demand are dim, largely due to continued population growth. In his 1995 book *Who Will Feed China?: Wake-up Call for a Small Planet*, Lester Brown documents how and why China will need to import more and more grain in the decades ahead in order to feed its expanding population. Brown notes that "[a]lthough the projections ... show China importing vast amounts, movements of grain on this scale are never likely to materialize simply because they, along with climbing import needs from other countries, will overwhelm the export capacity of the small handful of countries with an exportable surplus."

Add to this already grim picture the specter of oil depletion. It is not difficult to imagine the likely agricultural consequences of dramatic price hikes for the gasoline or diesel fuel used to run farm machinery or to transport food long distances, or for nitrogen fertilizers, pesticides, and herbicides made from oil and natural gas. The agricultural miracle of the 20th century may become the agricultural apocalypse of the 21st.

Expanding agricultural production, based on cheap energy resources, enabled the feeding of a global population that grew from 1.7 billion to over 6 billion in a single century. Cheap energy will soon be a thing of the past. How many people will post-industrial agriculture be able to support? This is an extremely important question, but one that is difficult to answer. A safe estimate would be this: *as many people as were supported before agriculture was industrialized*—that is, the population at the beginning of the 20th century, or somewhat fewer than 2 billion people.

There are those who argue that this figure is too low because new seed varieties and cultivation techniques developed during the past century should enable far more productivity per acre than farmers of the year 1900 were able to achieve.

This optimistic vision of the future of agriculture is currently being put forward by two camps with diametrically opposed sets of recommendations. One camp, consisting of the organic and ecological agriculture movements, recommends eliminating chemical inputs, shortening the distance between producer and consumer, and reducing or eliminating monocropping in order to support biodiversity. A recent report by Greenpeace International entitled *The Real Green Revolution: Organic and Agroecological Farming in the South* notes that in "this research we have found many examples where the adoption of [organic and ecological agriculture] has led to significantly increased yields."

The other camp, led by the agricultural biotechnology industry, has proposed an entirely different solution: the genetic engineering of new crop varieties that can outproduce old ones, grow in salty soil, or yield more nourishment than traditional varieties while requiring fewer chemical inputs. According to Hendrik Verfaille, President and CEO of Monsanto, the foremost corporate producer of gene-spliced agricultural seeds, this "technology increases ... crop yields, in some cases dramatically so. It is a technology that has been adopted by farmers faster than any other agricultural technology."

Optimists in both camps assume that energy conservation and alternative energy sources will cushion the impact of fossil-fuel depletion on agriculture.

But one could argue just as cogently that the figure of two billion as a long-term supportable human population is too high. Throughout the 20th century, croplands were degraded, traditional locally adapted seed varieties were lost, and farming skills were forgotten as the number of farmers as a percentage of the population—especially in industrialized countries—waned dramatically. These trends imply that, without fossil fuels, a smooth reversion to levels of productivity seen in the year 1900 may actually be unrealistically optimistic.

Organic or ecological agriculture can be even more productive in some situations than industrial agriculture, but local success stories cannot make up for the fact that the total amount of nitrogen available to crops globally has been vastly increased by the Haber-Bosch ammonia synthesis process, which is currently dependent on fossil fuels. Ammonia synthesis could be accomplished with hydrogen, which could in turn be produced with hydroelectic hydrolysis; but the infrastructure for such production is currently almost nonexistent. It will be extremely difficult to replace all or even a substantial fraction of the added available nitrogen from ammonia via organic sources (manures and legumes). John Jeavons, of the organization Ecology Action in Willits, California, has spent the past quarter century researching methods for growing a human diet on the minimum amount of land using no fossil-fuel inputs; he has concluded that survival is possible on as little as 2,800 square feet, enabling a theoretical maximum sustainable global carrying capacity of 7.5 billion humans. However, Jeavons' "biointensive" mini-farming method assumes the composting of all plant wastes and human wastes—including human bodies *post mortem*—and provides a strictly vegan diet with no oils and no plant

materials devoted to the making of fuels for cooking or heating. A more realistic post–fossil fuel carrying capacity would be substantially below the current population level.

As for the genetic engineering of food crops, the technology is risky and likely to have serious unintended environmental or health consequences that could more than wipe out whatever short-term benefits it may offer. Moreover it will not substantially reduce dependence on fossil fuels.

If we simply permit the optimistic and the pessimistic arguments to cancel one another out, at the end of the day we are still left with something like two billion as an educated guess for planet Earth's sustainable, long-term, post-petroleum carrying capacity for humans. This poses a serious problem, since there are currently nearly six-and-a-half billion of us, and our numbers are still growing. If this carrying-capacity estimate is close to being accurate, then the difference between it and the current population size represents the number by which human numbers will likely be reduced between now and the time when oil and natural gas run out. If that reduction does not take place through voluntary programs of birth control, then it will probably come about as a result of famines, plagues, and wars—the traditional means by which human populations have been culled when they temporarily surpassed the carrying capacity of their environments.

Heating and Cooling

Compared to food production, heating and cooling may seem far less consequential—matters merely of comfort. However, in many places—particularly the northern regions of North America, Europe, and Asia—a source of heat can mean the difference between life and death.

Currently in the US, according to the EIA, residential energy use accounts for 21 percent of the total national energy consumption. Of this, 51 percent is consumed for space heating, 19 percent for water heating, and 4 percent for air conditioning. The rest powers lights and appliances, including refrigerators.

Modern urban life offers a context in which it is easy to take heating and cooling for granted. Fuels and electrical power are piped or wired into houses and offices sight unseen and do their work silently and predictably at the turn of a .knob or the flick of a switch. Many office buildings have windows that cannot be opened, and few homes are designed for maximum energy efficiency. Temporary winter interruptions in fuel supplies often lead to deaths; and during the summer, elderly people who lack access to air conditioning are vulnerable to extreme heat. In an average year in the US, 770 people die from extreme cold and 380 from extreme heat; combined, these figures exceed the average combined death tolls from hurricanes, floods, tornadoes, and lightning. Serious and continuing fuel shortages would probably lead to a substantial increase in mortality from both temperature extremes.

Natural gas is widely used in industrialized countries for cooking and for heating hot water; diminishing supplies will obviously result in higher costs for these services. Only a relatively small proportion of the total amount of natural gas used goes toward cooking; however, since this is an essential function, a protracted interruption in supplies could have a major impact on people's daily lives.

Energy in the form of electricity is the primary power source for the refrigeration of food. As electricity becomes more expensive due to shortages of natural gas and the decline in net energy from coal, refrigeration will become more costly. Without refrigeration, supermarkets will be unable to keep frozen foods, and produce will remain fresh for much shorter time periods. The food systems of cities will need to adjust to these changes.

In areas of the world where wood, other plant materials, and dried animal wastes are used as fuel for space heating and cooking, air pollution and deforestation are already serious problems. Thus, given present population densities the substitution elsewhere of such traditional fuels for oil and natural gas will pose serious environmental and health hazards.

The Environment

The energy transition of the coming century will affect human society directly, but it will also likely have important indirect effects on the natural environment.

Some impacts—such as deforestation from increased firewood harvesting—are relatively easy to predict. As fossil fuels become scarce, it will become increasingly difficult to protect trees in old-growth forest preserves, and perhaps even those along the sides of city streets.

Other environmental effects of oil and natural gas depletion are less predictable. It is tempting to speculate about the impact on global warming, but no firm conclusions are possible. At first thought, it might seem that fossil-fuel depletion would actually improve the situation. With ever fewer gallons of gasoline and diesel fuel being burned in the engines of cars and trucks, less carbon dioxide will be released into the atmosphere to contribute to the greenhouse effect. Perhaps petroleum depletion could accomplish what the Kyoto protocols of greenhouse gas emissions have only begun to do.

However, it is important to remember that when global oil production peaks, half of nature's original endowment of crude will still be in the ground waiting to be pumped and burned. Extraction rates will gradually taper off but will not suddenly plummet. If efforts are made to increase coal usage in order to offset energy shortages from oil and natural gas, greenhouse gas emissions might remain close to current levels or even rise. Thus, unless a coordinate intelligent program is put in place for a transition to non-fossil energy sources as well as for a rapid and drastic curtailment of total energy usage, the net effect of oil and natural gas depletion on the problem of global warming is not likely to be significantly positive over the next few decades.

The situation is similar with regard to the problem of chemical pollution: a decline in the extraction of fossil fuels might seem to hold the promise of reducing environmental harms from synthetic chemicals. With less plastic being produced and fewer agricultural and industrial chemicals being used, the load of toxins on the environment should decrease. However, many pollution-monitoring, -control, and -reduction systems currently in place—including trash pick-up and recycling services—also require energy. Thus, even if the production of new chemicals declines, over the short run there may be heightened problems associated with the containment of existing pollution sources.

The reduced availability of oil and natural gas will likely provoke both electrical energy producers and politicians to call for a reduction of pollution controls on coal plants and for the building of new nuclear plants. But these strategies will entail serious environmental costs. Increased reliance on coal, and any relaxation on emissions controls, will result in more air pollution and more acid rain. And increased reliance on nuclear power will only exacerbate the unsolved problem of radioactive waste disposal.

As the global food system struggles to come to terms with the decline in available net energy for agriculture, transportation, and food storage, people who have the capacity to fish or to hunt wild animals will be motivated to do so at increasing rates. But given mounting energy and financial constraints, conservation agencies will find it difficult to control overfishing and the over-hunting of edible land animals. Endangered species will have fewer protections available and extinction rates will likely climb.

The environmental impacts of changing patterns in agriculture are difficult to predict, given that the direction of those changes is uncertain. If efforts are made to localize food production and to voluntarily reduce chemical and energy inputs via organic/eco-agricultural methods, then the current detrimental environmental impacts of agriculture could be reduced markedly. However, if the managers of global food systems opt for agricultural biotechnology and attempt to sustain inputs, negative environmental effects from food production are likely to continue and, in the worst case—a biotech "frankenfood" disaster—could be catastrophic.

National Politics and Social Movements

Democracy—the social means whereby citizens collectively and consciously control the conditions of their lives—is often regarded as an artifact of Greek civilization or the Enlightenment; but from a larger historical and anthropological perspective it can be seen as an attempt on the part of people living in modern complex societies to regain some of the autonomy and egalitarianism that characterized life in the hunter-gatherer bands of our distant ancestors. Democracy is a reaction against the concentrations of power that arose in early agricultural states and that burdened our more recent ancestors with kingship and serfdom. Because it implies that everyone should be able to participate in decisions regarding the allocation of resources, democracy is an inherently leftist ideal. This remains true despite the profoundly undemocratic nature of the Communist-bloc societies of the 20th century. Unquestionably, the most innovative thinking regarding democratic processes has come from the far-left of the political spectrum, which is occupied by anarchists of various stripes.

Both leftist and rightist ideologies contain an element of unreality or even denial concerning population and resource issues. Most rightists preach that all who wish for success and who work hard can potentially be wealthy if each individual is freed to compete in the market, unimpeded by government regulation. Most leftists promise that, if wealth is shared and decisions are made cooperatively, there will be plenty for everyone—with no exceptions in the face of population pressure or resource depletion. A few rightists acknowledge resource limits but argue that, since existence is a Darwinian struggle anyway, it is the fit (the wealthy) who should survive through economic competition while the unfit (the poor) are culled by starvation. A few leftists acknowledge limits but believe that, if humanity is made aware of them and empowered to deal with distribution issues democratically, people will decide to undertake a process of voluntary collective self-restriction that will enable everyone to, thrive within those limits. Typically, when either leftist or rightist regimes actually encounter resource

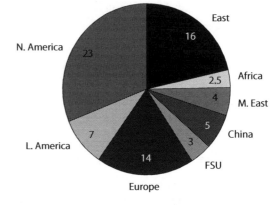

Figure 22a. Oil consumption by region, in millions of barrels per day. (Source: C. J. Campbell)

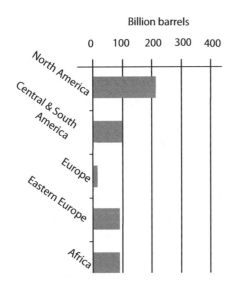

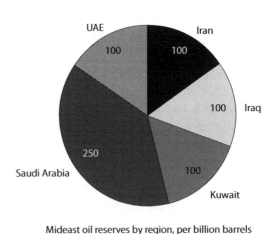

Mideast oil reserves by region, per billion barrels

Figure 22b. Oil reserves by nation Jan. 1, 2003. North American reserves include Canadian tar sands; Mideast figures are stated reserves, which are the subject of controversy. (Source: Oil & Gas Journal, 12/23/02)

limits, some aspect of ideology (democracy on the one hand, the free market on the other) is sacrificed, at least to some extent.

Because they have no solution, politicians on both sides will probably go to absurd lengths to obscure or mystify the real causes of the changes engulfing society. The public will likely not hear or read much about peaks in the extraction rates of oil or natural gas. They will see prices for basic commodities increase sharply (in inflation- or deflation-adjusted terms), but the ensuing economic turmoil will be held to be the fault of this or that social, political, ethnic, national, or religious group, rather than being identified as the unavoidable result of industrialism itself. The Left will blame selfish rich people and corporations; the Right will blame foreigners, "terrorists," and leftists.

Many people already sense that the traditional political categories of Left and Right no longer hold the solutions for today's unique social and environmental problems. Sociologist Paul Ray has argued, on the basis of extensive polling data, that a sizable portion of European and American populations consist of "cultural creatives" who defy both leftist and rightist stereotypes. These are people who typically espouse ecology and feminism while questioning globalization and the power of big business. It is conceivable that this constituency, if united and mobilized, could press for sensible energy policies.

The signal political development of the past decade has been the emergence of the global-justice movement advocating "globalization from below." That movement, to which many cultural creatives are drawn, demands the democratization of all social institutions and the limitation of the power of corporations to exploit workers in less-consuming countries; it also envisions a borderless world in which people can move without restriction. As the project of corporate globalization collapses for lack of energy resources, the anti-globalizationists

will see their warnings about the consequences of undermining local economies fully vindicated. However, corporate leaders may blame the global-justice movement for having helped cause the collapse of the global economy. And with dwindling resources motivating growing hordes to migrate en masse seeking necessities for survival, the ideal of a borderless world may seem less attractive to the settled segments of the populace.

The Geopolitics of Energy-Resource Competition

Though it is an empire in steep decline, the US—as the world's largest energy consumer, the center of the global industrial empire, and the holder of the most powerful store of weaponry in world history—will nevertheless play a pivotal role in shaping the geopolitics of at least the first decades of the new century. It is therefore probably best to begin an exploration of international relations during the net-energy decline with a survey of current US geopolitical strategy, especially as it relates to energy resources.

For the past few decades, the US has pursued a dual policy in the Middle East, the most oil-rich region of the planet. On the one hand, it has supported repressive Arab regimes in order to maintain access to petroleum reserves. America persuaded its Arab oil-state clients to denominate their production in US dollars. By thus being required to pay for most of their oil imports in dollars, importing countries around the world have contributed a subtle tithe to American banks and the US economy with every barrel of crude purchased. Arab rulers take a share of the "petrodollar" profits from the oil extracted from their countries and channel much of what they receive toward investments in the West and toward the purchase of US weapons. In exchange they have been promised US protection against their own people, who would naturally prefer to benefit more directly from the immense energy wealth with which nature has endowed their lands.

On the other hand, the US has supported Israel unquestioningly and with vast amounts of money and weaponry. Especially after its impressive military victory over the Arab states in 1967, Israel came to be seen as a foil to Arab nationalism. The Nixon Doctrine defined Israel's role as that of the "local cop on the beat," serving US military and intelligence interests in the region. This role became still more important following the Iranian Revolution in 1979, which denied the US its other main base in the Middle East. Israel also serves to deflect Arab resentment away from the United States: even though the US extracts considerable wealth from Arab countries, until recently Arab anger tended to be borne primarily by Israel, with the US depicting itself as a friend to the Arabs. American-backed Arab leaders likewise used Israel as a foil to divert the anger of the so-called "Arab street" away from their regimes' corruptness and toward the neighboring Jewish state.

Both long-standing US Middle-East policies are fraying. The two have had a history of mutual tension in any case; the most prominent instance of conflict between them was the (Arab-only) OPEC oil embargo against the West in late 1973 and early 1974, which was provoked by the US support for Israel during the Arab-Israeli war of 1973 and which resulted in severe economic hardship for the American economy.

Since September 11, 2001, the disharmony between these two policies has become truly cacophonous. The position of the rulers in Saudi Arabia and most of the small Gulf States is gradually being undermined, as these regimes have come under increasing public criticism in the US since September 11. The current Bush administration is no doubt reassuring these rulers that the US will continue to guarantee their security, but they must

nevertheless view the shift in American public opinion as worrisome. Their pariah status in the eyes of the American people in turn makes it more difficult for the monarchies to ignore the sentiment of their own populace—which is growing increasingly critical both of Israel's policies regarding the Palestinians, and of the US "war on terrorism." The invasion and ongoing occupation of Iraq add yet more division and uncertainty.

At the same time, the US leaders have announced to the world that they will unilaterally decide which actions around the world constitute "terrorism" and which do not. In the view of Arabs and Muslims everywhere, the US appears to have concluded that all actions by Palestinians against Israelis, whether against Israeli soldiers or against innocent civilians, constitute "terrorism." The US (and this includes most of the US media as well as the government) has also evidently adopted the attitude that no acts on the part of the Israeli government against Palestinians constitute "terrorism." Arabs view this as a double standard; for this reason as well, expressions of distrust and hatred of the US are mounting. When the Arab peoples see their own governments supporting America's self-defined "war on terrorism," their antagonism toward their, US-backed rulers intensifies. The rulers, seeing this deepening antagonism, are becoming increasingly uneasy.

It is impossible to say whether any of the governments in the oil-rich Arab nations whose stability and security the US has guaranteed for nearly 60 years will collapse in the near future. However, those governments are clearly now under greater internal stress than at any time in the past few decades. The Saudi royal family appears divided as to the line of succession from the ailing King Fahd. Moreover, most of the citizens of Saudi Arabia subsist largely on state subsidies derived from oil revenues, which have been falling partly due to population expansion. This fall in payments to the young can hardly help but cause social tensions within that country.

Thus, Saudi Arabia may well be headed toward turmoil, which could lead the US to intervene to seize the oil fields in the eastern part of that country. It seems at least possible that one of the purposes of the Iraq invasion was, in fact, to position new permanent military bases within easy striking distance of the Saudi fields.

Indeed, it may be that the disastrous outcome of the Iraq invasion has left the US with reduced, rather than expanded, options in the region.

While it is impossible to get inside the minds of US geopolitical strategists, statements by American officials suggest that they are at least contemplating the option of maintaining open supply lines through a variety of means tailored to the current realities in each of the world's oil-rich regions.

The Middle East: Next to the oil fields of Saudi Arabia, the world's largest petroleum reserves are those of Iraq, which were more or less withheld from the world market during the decade of sanctions, thus helping to keep global oil prices from falling too low.

Since the US-led invasion of Iraq, ongoing turmoil and outright sabotage have prevented the further development of that country's oil fields and export infrastructure to any appreciable degree. Indeed, in many months since the invasion, oil exports have lagged behind levels seen during the latter years of Saddam Hussein's regime.

To say that the Iraq occupation has not gone well is a serious understatement, and the consequences are likely to be grim. The US cannot simply leave, because to do so would create a power vacuum that could lead to political chaos throughout the region. But maintaining the present course will likely result in expanding resistance and civil war within the country, which could, in turn, lead to political chaos throughout the region. In short, it is difficult at this point to imagine a sequence of events leading to a peaceful and constructive outcome—barring some dramatic and unforeseeable change of strategy on the part of the US.

At the same time, regimes throughout the Middle East are on edge, seeking to rein in simmering anti-American sentiment among the Arab population, in order not to provoke further US overt or covert actions that would destabilize their governments.

Iran is likely to be a nexus of struggle in the near future. The US and Europe wish to deter Iran from developing nuclear weapons—which the Iranians see as essential to deterring American imperialist aggression. Meanwhile, both China and Russia are cooperating with Iran increasingly in the areas of energy and mutual defense. From a geopolitical perspective, Iran bridges the oil-rich regions of Middle East and Central Asia, lying adjacent to Iraq on the west and Afghanistan on the east. Iran is also a major oil and gas producer, and is thus crucial to the futures of importing nations. Moreover, the Iranian government has voiced interest in selling its energy resources in currencies other than the US dollar.

The Caspian Sea: Next to the Middle East, the Caspian Sea region contains perhaps the world's largest untapped reserves of both oil and natural gas (though these have probably been overestimated). Most of these reserves are in territory that was formerly part of the Soviet Union; some are bordered by Iran. In order to be marketable, these reserves must be accessible by pipeline. American geopolitical strategists are concerned not that these resources necessarily end up in the US, but that the US government and US-based corporations be in a position to control their flow, and hence their price.

This requires keeping any new pipelines from passing through Iran, which is still an oil power and is still operating independent of US control. American officials prefer an expensive route through Turkey to the Mediterranean, and another route through Afghanistan to Pakistan. In May 2005, the $4 billion Baku-Ceyhan-Tblisi pipeline—which begins in Azerbaijan and passes through Georgia and Turkey but bypasses Russia and Iran—opened, delivering one million barrels of crude per day to mostly European markets. Shortly after the recent US military action in Afghanistan, agreements were signed for the Afghan pipeline. However, despite backing from the World Bank and local governments, that project may be years from completion; current diplomatic efforts in that regard are essentially directed toward consolidating power in the region. This goal is also being sought through the same strategies used in the Middle East—by buying off corrupt regimes with promises of security and with shipments of arms for potential use against unruly civilian populations. Meanwhile the US has built 19 new military bases in the Caspian region, which appear to be permanent fixtures of the "war on terror."

Russia would prefer to see much of the Caspian's oil and gas flow north rather than south. Moreover, Russia has its own considerable oil and gas reserves and, even though its petroleum production peaked in 1987, it is in fact probably far better situated in that regard than the US over the short term. Currently it is exporting fossil fuels to Europe and the industrial nations of Asia. Throughout the 1990s, US leaders sought to use loans and debt to turn Russia into a dependent client state, and partly succeeded in that effort. However, Russian leaders are aware of the ace they hold in terms of their remnant military and industrial infrastructure, and their relatively abundant fossil-fuel reserves. Thus while the US and Russia remain overtly on friendly terms, the possibility of renewed geopolitical rivalry lurks close to the surface.

Under Vladimir Putin, Russia is seeking to regain some of the geopolitical prowess of the old Soviet Union. The privatization of industries and resources that occurred under Yeltsin has declined—as symbolized by the quashing of efforts by executives to sell Yukos (one of the largest Russian oil companies), to Western firms.

Russia's greatest advantage may lie simply in its geography: it is a vast country that covers much of the landmass of Eurasia.

If the US is to remain the world's superpower, it must dominate Eurasia, the site of two-thirds of the world's energy resources. This will be difficult to accomplish from thousands of miles away. America's oil imports must arrive by tanker, and this is an inherently vulnerable supply chain. The maintenance of imperial outposts likewise implies vulnerable supply chains stretching across oceans. In contrast, the countries of Eurasia can rely on pipelines, and on alliances based on geographic proximity. From a geostrategic point of view, an alliance between Russia, Europe, and perhaps China would be America's ultimate nightmare.

But this is exactly what is emerging, and the US has only itself to blame. The unilateralism of the Bush administration has predictably provoked collaborative activity on the part of countries that feel frozen out of decisions that affect their interests.

Barring an escalating confrontation over Iran, this geopolitical rivalry is not likely to erupt into a shooting war any time soon, but economic warfare seems nearly inevitable at this point. And here again, the US is extremely vulnerable, as concerted action by only a few nations could easily result in the severe undermining of the value of the US dollar.

South America: Venezuela is America's third largest oil supplier, and a prominent member of the Organization of Petroleum Exporting Countries (OPEC).

Soon after his election in 1998, Venezuela's president, Hugo Chavez, passed a spate of new laws that, among other things, increased the government's share of revenue from oil exports. Chavez also reformed Venezuela's constitution, through a constitutional assembly and a referendum, making it one of the most progressive constitutions in the world.

Given this record, the US-backed April 11, 2002, coup attempt against Chavez seemed wholly predictable. However, a successful counter-coup three days later reinstated Chavez, proving him to be a resourceful and resilient politician.

If Chavez sticks to his quasi-leftist principles, the US will likely search for other ways to reassert control over Venezuela's oil wealth. Further coup attempts are highly likely. Meanwhile, however, China is bidding for access to Venezuela's oil.

Meanwhile, in Colombia, the US has increased military aid to the regime of Alvaro Uribe Velez—ostensibly to help the Colombian army root out cocaine growers and smugglers. However, it is clear to nearly all international observers that another, perhaps more pressing goal is to secure US corporate interests—including oil fields, pipelines, and coal mines—from rebels in the country's 40-year-old civil war.[17]

China: The world's most populous nation possesses indigenous energy resources, but not on a scale large enough to fuel its accelerating process of industrialization. Continued reliance on domestic coal supplies has economic advantages, but it will entail environmental devastation and will be incapable of powering the development of China's transportation infrastructure. With its burgeoning appetite for energy, China is capable of dramatically changing the global supply/demand picture for oil and natural gas. Until recently, the US provided the marginal demand in crude oil. But now China is building refineries at a rapid rate, even as its consumption of crude far outpaces its indigenous production. China is using Dickensian sweat shops and near-slave labor in order to grow its economy; but its leaders know that, in order for its efforts at industrialization to succeed, human labor must increasingly be tied to fuel-fed machinery.

China has recently surpassed Japan to become the world's second foremost oil importer (the US is still first in line, importing twice as much as China and Japan combined). Increasingly, China and the US are competing for long-term oil export contracts in Central Asia, the Middle East, Africa, and even Canada and South America.

China's economic influence is expanding quickly throughout Asia—including the contested Caspian Sea region—bringing it inevitably into conflict with US strategic interests there. Here as elsewhere, American strategists would prefer to avoid direct confrontation, as China's increasing share of the global economy and its massive production of export goods for the US market ensure that any open conflict would inevitably harm both sides. Nevertheless, since China is capable of absorbing a quickly growing share of the available global oil exports, economic and possibly military conflict with the US is likely sooner or later.

Economic warfare between the two nations would damage both severely. The US has been able to run up massive deficits in recent years partly because of China's willingness to purchase American government debt in the form of Treasury Bills. China could thus help precipitate a collapse of the US dollar merely by dumping its investments on the international market. However, this would hurt China as well, since that country is dependent on food imports from the US, which could be halted if competition turns ugly.

China also has strategic energy-resource interests in the South China Sea that overlap with those of nations other than the US. The area—bordered on the north by China, on the east by the Philippines, on the south by Indonesia and Malaysia, and on the west by Vietnam—is believed to possess significant undersea resources of gas and oil (though exploration efforts to date have been disappointing). All of the nations in the region have conflicting claims on those resources. As policy analyst Michael Klare has pointed out in his book *Resource Wars: The New Landscape of Global Conflict,* growing demand for energy in Asia will affect the South China Sea in two significant ways. First, the states that border on the area will undoubtedly seek to maximize their access to its undersea resources in order to diminish their reliance on imports. Second, several other East Asian countries, including Japan and South Korea, are vitally dependent on energy supplies located elsewhere, almost all of which must travel by ship through the South China Sea. Those states will naturally seek to prevent any threat to the continued flow of resources. Together, these factors have made the South China Sea the fulcrum of energy competition in the Asia-Pacific region.

In recent years, China has seized several islands from Vietnam and established military outposts on them; meanwhile, most of the nations in the region have embarked on an arms race to protect shipping lanes and defend resource claims.

Britain: Only a few years ago British Prime Minister Tony Blair was hailing the new information economy as a replacement for the old oil economy. Then came the oil price spike of 2000, which wreaked temporary havoc on London's financial markets. Perhaps Blair has since come to appreciate the significance of the fact that the rate of his nation's share of the North Sea oil and gas extraction appears to have peaked in 1999–2000. North Sea oil gave the UK a tremendous economic boost during the past three decades; but as of 2005 Britain has ceased to be an oil exporter and will need to import increasing amounts of petroleum in coming years in order to maintain its economy. British coal production is also in steep decline. Blair may have decided that his nation's economic survival hinges on future access to the resources of the Middle East and that the best way to ensure that

access is through maintaining a close military and political alliance with the US. Though his positions on issues in this regard are often unpopular among his constituents, Blair is forced by circumstances to provide the US with aid and cover in its otherwise unilateralist pursuit of global resource dominance.

The Balkans: This is not a resource-rich region, but one essential to the transfer of energy resources from Central Asia to Europe. It is also the site of Camp Bondsteel, the largest "from-scratch" foreign US military base constructed since the Vietnam War. Located in the Yugoslav province of Kosovo on farmland seized by US forces in 1999, Camp Bondsteel lies close to the US-sponsored Trans-Balkan oil pipeline, which is now under construction. Brown and Root Services, a Houston-based contractor that is part of the Halliburton Corporation, the world's largest supplier of products and services to the oil industry, provides all of the support services to Camp Bondsteel—including water, electricity, spare parts, meals, laundry, and fire fighting services.[20]

While it would no doubt be an oversimplification to say that US military action in the Balkans in the 1990s was motivated solely by energy-resource considerations, it might be just as wrong to assume that such considerations played only a minor role.

Regional rivalries and long-term strategy: Even without competition for energy resources, the world is full of conflict and animosity. For the most part, it is in the United States' interest to prevent open confrontation between regional rivals, such as India and Pakistan, Israel and Syria, and North and South Korea. However, resource competition will only worsen existing enmities.

As the petroleum production peak approaches, the US will likely make efforts to take more direct control of energy resources in Saudi Arabia, Iran, the Caspian Sea, Africa and South America—efforts that may incite other nations to form alliances to curb US ambitions.

Within only a few years, OPEC countries will have control over virtually all of the exportable surplus oil in the world (with the exception of Russia's petroleum, the production of which may reach a second peak in 2010, following an initial peak that precipitated the collapse of the USSR). The US—whose global hegemony has seemed so complete for the past dozen years—will suffer an increasing decline in global influence, which no amount of saber rattling or bombing of "terrorist" countries will be able to reverse. Awash in debt, dependent on imports, mired in corruption, its military increasingly overextended, the USA is well into its imperial twilight years.

Meanwhile, whichever nations seek to keep their resources out of the global market will be demonized. This has already occurred in the cases of Iran, Iraq, and Libya—which sought to retain too large a share of their resource profits to benefit their own regimes and hence attained pariah status in the eyes of the US government. Essentially they were seeking to do something similar to what the American colonists did in throwing off British rule over two centuries earlier. Like the American colonists, they wanted to control their own natural resources and the profits accruing from them. Many readers will object to such an analogy between American colonial patriots and modern-day Libyan or Iraqi leaders on the grounds that the latter are, or were, autocrats guilty of human-rights abuses that justified their condemnation by the international community. But we must recall that America's founders were themselves engaged in slavery and genocide and that many US client states—including Turkey, Israel, Indonesia, and Saudi Arabia—have also been guilty of serious abuses.

In the future, secure access to resources will depend not only on the direct control of oil fields and pipelines but also on successful competition with other bidders for available supplies. Eventually, the US will need to curtail European and Japanese access to resources wherever possible. Again, every effort will be made to avoid direct confrontation because in open conflict all sides will lose. Even the closest trading partners of the US—Canada and Mexico, which are currently major energy-resource suppliers—will become competitors for their own resources when depletion reaches a point where those nations find it hard to maintain exports to their energy-hungry neighbor and still provide for the needs of their own people.

Civil wars will be likely to erupt in the less-industrialized nations that have abundant, valuable, and accessible resources, such as oil, natural gas, and diamonds, rather than in those that are resource-poor. This conclusion is based on a correlation study by Indra de Soysa of the University of Bonn of the value of natural resources in 139 countries and the frequency of civil wars since 1990. The finding runs counter to the long-held assumption that internecine warfare is most likely to occur in resource-poor countries. Often rival groups within nonindustrial countries use wealth from the sale of resources—or from leases to foreign corporations to exploit resources—in order to finance armed struggles. Pity the nations with resources remaining.

The least industrialized of the world's nations will face extraordinary challenges in the decades ahead, but may also enjoy certain advantages. Industrialized nations will seek to choke off the flow of energy supplies to resource-poor economies, most likely by yanking their debt chains and enforcing still more structural-adjustment policies. However, less-industrialized nations are able to squeeze much more productivity out of energy resources than are the energy-saturated economies of the industrialized nations. Less-industrialized nations are therefore potentially able to bid prices higher, or to absorb higher energy costs much faster, than the industrialized nations. This is only one of many wild cards in the longer-term game that will be played out as the world's energy resources slowly dribble away.

Managing the Collapse

Virtually all of the authors who have contributed to the literature on sustainability tell us that, in order for a transition to a lower-complexity and lower-throughput society to occur without a chaotic collapse, humanity will have to take a systemic approach to resource management and population reduction.

In this final chapter I intend to sketch the general outlines of the social economic, political, and individual-lifestyle changes that are needed in order to minimize the consequences of energy-resource depletion and to build the foundations of a society capable of enduring for many generations into the future. I will save two important questions—*Is it too late?* and, *Are these recommendations realistic?*—until the end of the chapter.

You, Your Home, and Your Family

There is much that you, as an individual, can do to prepare for the energy transition. Below are suggestions grouped into eight categories; but as you take your first steps on the path toward a sustainable lifestyle, you will find that these strategies naturally blend into each other. You will also find new friends who are on the same path and who can offer encouragement and suggestions. Many thousands of people find satisfaction in making these sorts of efforts for their own sake—not just as a strategy for survival during an anticipated social or economic crisis. Over the past decade or so, my wife and I have employed most of these strategies in our home and with our community of friends; and my colleagues and students and I explore them in some detail in our yearlong program on Culture, Ecology, and Sustainable Community at New College of California.

Energy usage. Begin by assessing your current energy usage, then decide which areas of usage are essential and which are nonessential. Gradually and deliberately reduce your nonessential usage. This is a process that may continue over some time and may require considerable experimentation and ingenuity. Examine your utility bills carefully and begin using them as a feedback mechanism to tell you how you are doing in your conservation efforts.

You can improve the energy efficiency of your home relatively easily by replacing incandescent lights with compact fluorescents; by more thoroughly insulating walls and roof; by replacing single-pane glazing with high-e double-pane windows; and by choosing energy-thrifty appliances. Direct most of your effort toward the area where your energy usage is greatest. For most people, this will be home heating.

Alternative energies. After you have pared your energy usage to the bare minimum, consider equipping your home with a renewable energy source. Photovoltaic systems are expensive now, but when electricity prices begin to soar you might be glad you invested in one.

Wind power may be feasible for you if you live in a rural setting. Small wind turbines generate power more efficiently than do PV panels, but they require tall towers and can make an unpleasant noise.

If you rent your house or apartment, altering the building itself may seem unfeasible. You may instead wish to examine your housing options: might it make sense to move to a place where it would be easier to pursue radical energy efficiency and energy independence?

Your home. If you are thinking of building a new home or remodeling your existing one, consider using ecological design principles and natural or recycled materials. Straw-bale, rammed-earth, and cob construction can be used to build houses that stay warm in the winter and cool in the summer with little or no energy usage. Many counties now routinely grant building permits for these kinds of alternative structures, which have proven themselves over time to be durable and efficient.

Building one's own structure is an extraordinarily empowering experience. If you lack construction skills, take a workshop on basic carpentry and find a builder who is familiar with natural building, who is willing to teach you, and who will allow you to do as much of the work as you can.

If you live in a rural or semi-rural area, a composting toilet might be a good alternative to a conventional septic system, in that it would allow you to use human wastes as fertilizer for trees and shrubs. As discussed by Joe Jenkins in his *Humanure Handbook,* there are even simpler and more direct methods for composting human waste, though local ordinances typically prohibit them.

Finances. Reduce your debt. Whatever interest you are paying on loans—especially credit-card interest—is nonproductive and a drain on your personal energy budget. Further, don't buy what you don't absolutely need. Forget about your "patriotic duty" to the consumer economy and to the maintenance of the national financial system. Your primary duty is to a higher cause: personal and planetary survival.

Exiting the consumer treadmill is psychologically as well as financially freeing. The "voluntary simplicity" movement has been growing internationally for the past two decades, and local networks and support groups exist in many areas.

Appropriate technology. Begin replacing some of the Class D tools in your life with Class A, B, and C tools (see Chapter 1, pp. 25, 26). A well-made hand tool—a hoe, garden spade, saw, chisel, or plane—is a joy to use; employing it properly requires skill, but offers considerable satisfaction.

These days it is often more time-consuming and expensive to repair and reuse manufactured objects than simply to throw them away and replace them. But as energy resources become more scarce and valuable, having basic maintenance and repair skills could mean the difference between continual

frustration and lack on the one hand and sufficiency and satisfaction on the other. Many junior colleges offer classes to the public on small-motor repair. Knowing how some of the simple devices we depend on actually operate tends to raise one's level of self-confidence, even in the absence of energy shortages. Begin to assemble a small library of books and articles on home repair and maintenance and begin to try fixing simple things on your own, seeking advice whenever necessary.

Health care. Perhaps the most important appropriate technologies are those for health maintenance. Learn about healing herbs and basic medical procedures that can save lives in the temporary absence of doctors and hospitals. Start a medicinal herb garden in your back yard or window box and assemble a natural home medicine chest consisting of dried herbs and herbal tinctures, as well as books on natural first-aid remedies.

Food. Grow as much of your own food as you can. Doing so successfully will require practice and experimentation: gardening is both an art and a science. If you live in an apartment, explore window-box or hydroponic gardening.

Unless you have a very large city lot or some acreage, considerable gardening experience, and a fair amount of time on your hands, it will be unrealistic to expect to grow all of the food you will need to sustain yourself and your family However, you can make it your goal to grow more of your diet each season by managing your garden more carefully, and by planting a wide variety of vegetables that can be harvested more or less continuously A greenhouse or cold frame can help extend your growing season year-round.

Saving seed is a time-honored traditional craft that contributes both to self-reliance and to the maintenance of the genetic commons. Buy open-pollinated, non-hybrid varieties of vegetable seeds, and in your garden set aside some space where a few plants from each variety can complete their life cycles, yielding seeds for next year. Seek out neighbors who are avid gardeners and whose families have lived in your area for several generations: they may have heirloom seed varieties, well adapted to your local soil and climate, that they would be happy to share with you in return for some of your own more unusual seeds or produce.

Look for alternatives to chemical fertilizers, pesticides, and herbicides. Some nurseries specialize in supplies for the organic gardener.

In order to keep from quickly depleting your soil, you will need to build and renew it each year. You can make your own compost from lawn clippings, leaves, soil, manure, kitchen scraps, and crops grown especially for the compost pile. A worm box—which turns kitchen scraps into rich black humus—can be employed even in a small apartment.

If you have the space, keeping a few chickens can serve several purposes at once: chickens can produce both food (eggs and, if you wish, meat) and nitrogen-rich fertilizer while periodically ridding your garden of snails, slugs, and invasive insects.

Food self-reliance entails devoting some thought and effort toward preservation and storage. Drying is the easiest means of preservation, and it requires no energy source other than the Sun. Canning takes more planning, work, and energy, but enables you to put up larger quantities.

It is easy to construct a solar oven that will cook food even on cold winter days, as long as the Sun is out. A meal consisting entirely of food that you have grown and that you have cooked slowly in a solar oven is truly a banquet of self-sufficiency.

Transportation. For most Americans, the automobile is a key personal link to the oil-based energy regime of industrialism. A new car rolls off the assembly line each second, and the global fleet uses twice as much energy from oil as humans obtain from the food they eat.

Consider the possibility of living car-free. When Katie Alvord, author of *Divorce Tour Car!*, went on a national book tour, she traveled by bus, tram, and folding bicycle just to prove that, even in our auto-dependent society, it can be done. As a first step, go car-free one day a month, then one day per week. Drive only when necessary and walk, use mass transit, or carpool whenever possible. Vancouver and San Francisco now have car co-ops: if you live in either of these cities, join one; if you don't, start one. If you must have a personal vehicle, give some thought to the kind of car you drive. When it comes time to replace your metal monster, consider the alternatives: buying an older used car will entail a smaller energy cost than buying a new one, and buying a gas-thrifty model may save both fuel and money. Electric and hybrid cars are now available, and it is possible to operate a diesel vehicle on biodiesel fuel—or, after a small alteration to the fuel system, to run it on recycled vegetable oil that can be obtained at little or no cost.

Your Community

The strategy of individualist survivalism will likely offer only temporary and uncertain refuge during the energy downslope. True individual and family security will come only with community solidarity and interdependence. If you live in a community that is weathering the energy downslope well, your personal chances of surviving and prospering will be greatly enhanced, regardless of the degree of your personal efforts at stockpiling tools or growing food.

During the energy upslope, most traditional communities became atomized as families were torn from rural subsistence farms and villages and swept up in the competitive anonymity of industrial cities. Security during the coming energy transition will require finding ways to reverse that trend.

The first steps can be taken immediately. Get to know your neighbors. Look for people in your community who share a similar interest in voluntary simplicity and self-reliance, and begin to form friendships and habits of mutual aid.

Your community may be a village, a neighborhood, or a city. Regardless of its size, you will find opportunities for building alliances. There are inevitable challenges inherent in this project: your community no doubt includes people whose interests conflict with your own. Rather than identifying "wedge" issues and highlighting disagreements, find areas of common interest that are related to the goal of community sustainability and security regarding universal human needs: food, water, and energy.

The Post Carbon Institute (www.postcarbon.org) has prepared materials to assist with creating community responses to peak oil (see sidebar p. 240). Consider starting a Post Carbon "outpost" in your town or city. Begin by hosting community meetings on the subject, then organize action committees.

Food. Few communities bother to examine the security and sustainability of the food system on which they rely. Berkeley, California, is one of the first cities to deliberately undertake such an assessment. In 2001, the Berkeley City Council passed the Berkeley Food and Nutrition Policy, which provides the community with a clear framework for the next decade to help guide its creation of a system that, in the words of its mission statement, is "based on sustainable regional agriculture," that "fosters the local economy," and that "assures all people of Berkeley have access to healthy, affordable and culturally appropriate food from non-emergency sources." Berkeley's schools have adopted a policy of serving organic food to students and identified the goal of having a garden in each school.

Community gardens provide good food and build community at the same time by transforming empty lots into green, living spaces. Members of the community share in both the maintenance and rewards of the garden. Today there are an estimated 10,000 community gardens in US cities alone.

A particularly innovative community gardening project was begun in Sonoma County, California, in 1999 by a group of idealistic college students. Calling their loosely knit organization Planting Earth Activation (PEA), the leaderless collective offers to dig and plant gardens for anyone free of charge; all they ask in return is a share of open-pollinated seed saved from those gardens, with which they can plant still more gardens. Garden plantings occur in a party atmosphere, with music and food to accompany the hard work. PEA chapters have recently sprung up in neighboring counties as well.

Community-supported agriculture (CSA) is a fast-growing movement that operates on the premise that the consumer contracts directly with a farmer. The CSA model of local food systems began 30 years ago in Japan, where a group of women concerned about the increase in food imports and the corresponding decrease in the farming population initiated a direct purchasing relationship between their group and local farmers. The concept traveled first to Europe and then to the US. As of January 2000, there were over 1,000 CSA farms across the US and Canada. CSA members cover a farm's yearly operating budget (for seeds, fertilizer, water, equipment maintenance, labor, etc.) by purchasing a share of the season's harvest, thus directly assuming a portion of the farm's costs and risks. In return, the farm provides fresh produce throughout the growing season.

Water. Unless you have a well or live next to a stream or lake, access to water is more of a community issue for you than an individual one. And since water inevitably flows—both above and below the ground surface—all water issues are ultimately community issues.

Because water treatment plants and pumping stations use energy, communities will need to conserve water and find new ways to distribute water and prevent water pollution as energy resources become more precious.

Some communities have already made some efforts along these lines. After suffering through years of drought, Santa Barbara County, California, instituted a Water Efficiency Program, which offers information on home and landscape water conservation as well as educational materials on water conservation. And a grassroots group in Atlanta, Georgia, has initiated Project Harambee: they distribute free ultra-low-flush toilets, low-flow shower heads, and energy conservation information to low-income households in an effort to reduce water and energy consumption. There are similar programs in dozens of other towns and cities across North America.

Watershed protection groups and Water Watch programs also exist in regions throughout the US, monitoring rivers and streams and identifying sources of pollution. Members collect data, which they share with county and state agencies, and educate the public through literature, classes, and tours.

Natural waste-water treatment facilities, which rely on the purifying characteristics of marsh plants, are operating successfully in Germany, Switzerland, and the Netherlands; in the US, a natural waste-water treatment facility began operations in Arcata, California, in 1986. Arcata's project uses a marsh system to provide both secondary treatment for the city's waste water and wildlife habitat.

Despite initial steps like these, few communities are prepared to meet energy-based challenges to their ability to supply clean water to citizens. The development of alternative low-energy water delivery and treatment systems in large and small communities everywhere will require creativity and cooperative effort. As communities begin to prioritize their energy budgets, they will need to devote whatever power they obtain from renewable sources, such as photovoltaic and wind, first toward

water, as a foundation for their collective survival and sustainability.

Local economy. Corporate globalization has hit local economies hard. In town after town, local businesses have succumbed to "big box" chains like Wal-Mart, which buy in huge quantities and often sell mostly imported items made by low-paid workers. Once a local economy has been destroyed by dependence on the "big box," the chain frequently pulls out, forcing members of the community to drive tens of miles to the nearest larger town for basic consumer needs. Several communities have successfully resisted Wal-Mart; their campaigns have relied on group initiative, hard work, and hired consultants.

Resistance to chains must be accompanied by efforts to promote and sustain local enterprises: locally owned bookstores, restaurants, grocery stores, clothing stores, and product manufacturers. "Buy local" advertising campaigns can-help keep regional economic infrastructures robust, enabling them better to face both the immediate threat of competition from national chains and the approaching challenge of the energy transition.

Since national currencies are based on debt, their use subtly but inevitably saps wealth from local communities. Every dollar loaned into existence requires the payment of interest, some of which (even if the loan is issued by a local credit union) goes to a nationally chartered banking cartel. Debt-based money thus systematically transfers wealth from the poor to the rich. In addition, national currencies are subject to inflation, deflation, and collapse as well as to manipulations and panics beyond the community's control. One solution is the promotion of local barter systems; another is the creation of a local currency. Both are legal (if operated within certain guidelines) and have long histories of success.

Public power. In many cities, electricity and natural gas are delivered by publicly owned not-for-profit power utilities, as opposed to investor-owned utility companies. There are currently roughly 2,000 municipal power districts in the US, which together deliver electricity to 15 percent of the population.

Public power enables every citizen to be a utility owner, with a direct say in policies that affect not only rates and service but choices as to energy sources as well. Citizens can, for example, decide to phase out nuclear plants and replace them with wind or solar plants—as happened in the case of the Sacramento, California, Municipal Utility District (SMUD).

Starting a publicly owned power electric utility takes considerable time, money, and effort. Nevertheless, currently several large cities and many small towns in all parts of the US are considering establishing their own utilities in order to save money and provide citizens with more control. More than 40 public power utilities have been formed in the last two decades.

The necessary steps in forming a public power utility vary from state to state, but typically include authorizing a feasibility study; analyzing pertinent local, state, and federal laws; obtaining financing; informing and involving the public; holding an election to let voters decide on the merit of the proposal; and issuing bonds to buy present facilities or finance the construction of new ones.

Citizens in many states have formed energy co-ops, which buy electricity in quantity and sell to members at a discount. Energy co-ops are private, independent electric utility businesses owned by the-consumers they serve. Distribution cooperatives deliver electricity to the consumer, while generation-and-transmission cooperatives (G&Ts) generate and transmit electricity to distribution co-ops. Currently in the US, 866 distribution and 64 G&T cooperatives serve 35 million people in 46 states. In distribution co-ops, members can decide what sources to buy from: coal, nuclear power, or renewables; G&T co-op members can choose what

kinds of plants to invest in. Initial investment money can often be obtained from a local credit union.

Community design. Towns and cities are continually changing, and most communities have some process in place to plan their future direction of change. When citizens become involved in the urban planning process and bring with them the values of sustainability and conservation, important strides can be made toward successfully weathering the energy transition.

During the last century, most towns and cities grew around the priorities of the automobile. Today, it is essential that communities be redesigned around people. Public transportation, walking, and bicycling must be emphasized—and car traffic discouraged. A first step is the creation of car-free zones in mid-town areas. Often such zones are a boon to local businesses; and as anyone knows who has visited old European cities like Venice or Siena, a car-free town or town center offers far more space for cultural expression than is possible in a car-dominated city. Neighborhoods can be made more pedestrian-friendly with speed bumps, snaky curves in roads, and prominent crosswalks. And towns can systematically reduce automobile traffic with carpool-only lanes. Meanwhile, funds can be diverted away from road building and toward the provision of light-rail service.

Cities and towns can also be encouraged to build more bike paths and bike lanes. Some cities are already far ahead of others in this regard. Portland, Oregon, for example, has a fleet of refurbished old bicycles available for free use within the city, showers and changing rooms for bike commuters, buses and trains that accept bikes on board, and a 140-mile bike trail encircling the city.

Usually, urban design priorities like these are articulated and promoted by citizens' advocacy groups. Concerned citizens have established sustainability groups in several cities and counties (including Sustainable Seattle, Sustainable San

Francisco, and Sustainable Sonoma County); they perform sustainability studies using the Ecological Footprint indicator and develop and advocate plans to improve the community's environmental health, to stabilize its local economy, and to achieve a more equitable distribution of resources. Sustainability groups examine issues of air quality, food, hazardous materials, waste, water, biodiversity, parks, and open spaces; their working groups include representatives from city agencies, local businesses, and academia.

The "new urbanism" movement, discussed in the documentary "The End of Suburbia," advocates making cities more pedestrian friendly by building up multi-use urban cores and neighborhoods and discouraging strip-mall corridors. A more radical approach is advocated by architect Richard Register, author of *Ecocities: Building Cities in Balance with Nature* (Berkeley Hills, 2001), who understands the challenge of the imminent oil production peak and envisions redesigning cities to virtually exclude the personal automobile.

Local governance. One way to make change within your community is to get involved in local politics. Politics is about power and decision-making, and it entails conflict and hard work. When the stakes are high, the political process almost inevitably becomes subject to corruption, and cynicism and burnout usually follow. However, in smaller communities, local government is still relatively open to citizens' input and participation.

Involvement in local politics opens many possibilities for moving your community in the direction of sustainability. Find out what local issues are. Go to meetings of your city council or board of supervisors. Identify people and groups with concerns similar to yours and work with them to form research and action committees.

It is possible to influence officeholders by writing letters, actively participating at public meetings, or writing opinion pieces for the local newspaper. It is also possible to run for office—even though it takes a certain kind of personality to want to do so. This requires knocking on doors, but it also presents an opportunity to educate the public.

Intentional communities. For many centuries, idealists have sought to create a better world by building model communities in which alternative ways of living can be experimented with and demonstrated. There are thousands of intentional communities in existence today, and others in the formative stages, many of which are pioneering a post-industrial lifestyle.

During the energy decline, life in an intentional community could offer many advantages. Association with like-minded people in a context of mutual aid could help overcome many of the challenges that will arise as the larger society undergoes turmoil and reorganization. Moreover, new cooperative, low-energy ways of living can be implemented now, without having to wait for a majority of people in the larger society to awaken to the necessity for change.

However, these advantages do not come without a price: to live successfully in an intentional community requires work and commitment. Many communities fail and many members drop out, for a variety of reasons—often centering on the individuals' projection of unrealistic expectations onto the group. Nevertheless, some communities have managed to survive for decades, and few members of successful communities would willingly trade their way of life for that of the alienated urbanite.

Ecovillages are urban or rural communities that strive to integrate a supportive social environment with a low-impact way of life. They typically experiment with ecological design, permaculture, natural building materials, consensus decision-making, and alternative energy production. Ecovillages currently scattered around the world include the Findhorn Foundation in Scotland; Eco Village at Ithaca, in Ithaca, New York; the Farm in Summertown, Tennessee; Earthaven in North Carolina; and Mitraniketan in Kerala, India. One that I have visited, the Dancing Rabbit Ecovillage in northeastern Missouri, was established in the early 1990s by a group of young West Coast recent college graduates. Today the community consists of about fifteen adults and children. The buildings were constructed by the residents from natural materials (straw bales and earth), their vehicles all run on vegetable oil, and the community has its own currency and grows most of its food on-site.

The Nation

It is easiest to exert influence on the political process at the local community level. State and provincial politics are almost invariably subject to more competition for access and power, and thus to more corruption. At the national level, the degrees of competition and corruption are truly daunting.

However, many of the legal and economic structures that prevent industrial societies from more quickly and more easily adapting to the energy decline can only be altered or replaced nationally. Even small, incremental changes at this level of government can have important effects. Thus it is essential for citizens who are aware of energy-resource issues to direct at least some of their efforts toward encouraging change at the highest levels of political organization.

Alternative energies and conservation. As nonrenewable energy sources become depleted, it is crucial that renewable substitutes be developed and implemented to replace them—to the degree that replacement is possible. Individuals and local communities can help this happen, but a systematic national policy is badly needed. Policymakers cannot simply wait for the price of nonrenewable sources to rise and that of renewables to fall so that the market itself automatically effects the transition. It will take decades to rebuild the national energy infrastructure, and price signals from the dwindling of nonrenewables will appear far too late to be of any help. In fact, it is already too late to make the transition painlessly. An easy transition might have been possible if the nation had begun the project in the 1970s and continued it consistently and vigorously through to the present. Still, even at this late date, a truly heroic national effort toward developing renewables could succeed in substantially reducing social chaos and human suffering in the decades ahead.

Until recently, the US was spending more on renewable-energy technologies than any other nation, on both a per-capita and an absolute basis. However, Germany, Japan, Spain, Iceland, and the Netherlands are now moving quickly ahead with renewables, while national efforts in the US are stalled.

In view of the absolute dependence of industrial society on energy resources and the imminent decline in fossil fuel availability, one would think that the search for alternatives must be the nation's first priority. Yet the 2004 budget gives its biggest priority instead to the military. The Bush administration's proposed military budget *increase* from 2003 to 2004 is itself larger than the *entire* military budget of any other country in the world except Russia. One cannot help but wonder: Which would be more likely to provide security for us and the next generation—yet another expensive weapons system or a reliable, non-depletable source of energy?

The federal government could and should speed the transition to renewable energy sources by providing substantial tax breaks for individuals who invest in wind and solar, and subsidies for utilities that switch to renewables. Carbon taxes should be implemented and gradually raised—not only to discourage the use of nonrenewables, but also to provide funds for rebuilding the energy infrastructure.

Conservation should also be a high priority: the inability of Congress to pass laws mandating higher auto fuel-efficiency standards is an embarrassment and a disgrace. However, stringent efficiency mandates should be passed not only for automobiles but for a range of appliances and industrial processes. The nation should set goals of reducing the total energy usage by two percent per year, and of progressively altering the ratio of nonrenewable to renewable sources. There may be an economic price for such policies, but it will pale in comparison to the eventual costs of the present course of action.

Food systems. We need to redesign our national food system from the ground up. Currently, that system is centralized around giant agribusiness corporations that control seeds, chemicals, processing, and distribution. Most farmers are economically endangered. We need a national food policy that encourages regional self-sufficiency. This will require a 180-degree shift in how farm subsidies are designed and applied.

Current farm subsidies encourage huge agribusiness corporations and energy dependence. As Philip Lee recently argued in an article in the *Ottawa Citizen*, Canada's agricultural subsidies are similarly promoting centralized, fuel-fed agriculture over sustainable, diversified, local food production.

A range of problems surrounding industrial agriculture could be solved simply by ending current farm subsidies—or, better yet, by instituting an entirely different regime of subsidies that would benefit diversity rather than monocropping; small family farms rather than agribusiness cartels; and organic farming rather than biotech- and petro-chemical-based farming.

Financial and business systems. The changes needed in the national economic structure go far beyond efforts to improve accounting regulations so as to avoid more corporate bankruptcies on the scale of Enron or WorldCom. The entire system—designed for an environment of perpetual growth—requires a complete overhaul.

Giant corporations are engines of growth and have become primary power wielders in modern industrial societies. One way to rein them in would be to challenge important legal privileges they have acquired through dubious means. The Fourteenth Amendment to the US Constitution was adopted soon after the Civil War to grant freed slaves the rights of persons; but by the last decades of the 19th century, judges and corporate lawyers had twisted the Amendment's interpretation to regard corporations as persons, thus granting them the same rights as flesh-and-blood human beings. Since then, the Fourteenth Amendment has been invoked to protect corporations' rights roughly 100 times more frequently than African Americans' rights.

The legal fiction of corporate personhood gives corporations the right of free speech, under the First Amendment to the US Constitution. In recent years, when communities or states have sought to restrict corporations' campaign donations to politicians, the courts have overruled such restrictions as a violation of corporate free-speech rights as persons. Corporations also are allowed constitutional protection against illegal search and seizure so that decisions made in corporate boardrooms are protected from public scrutiny. However, corporate "persons" do not have the same limitations and liabilities as flesh-and-blood persons. A human person in California who commits three felonies will be jailed for 25 years to life under that state's "three-strikes" law; but a California-chartered corporate "person" that racks up dozens of felony convictions for breaking environmental or other laws receives only a fine, which it can write off as the cost of doing business. Personhood almost always serves the interests of the largest and wealthiest corporations while small, local businesses that also have corporate legal status are systematically disadvantaged.

Americans should unite behind a national movement to rescind corporate personhood; until that goal is achieved, they should petition state attorneys general to review or revoke the charters of corporations that repeatedly harm their communities or investors.

Our current monetary system, which is based on debt and interest and thereby entails endless economic growth and snowballing indebtedness, requires replacement. While some monetary theorists advocate a gold-based currency as a solution, others argue that a well-regulated, non-debt-based paper or computer-credit currency would have greater flexibility. There is at least one precedent in this regard: the Isle of Guernsey, a British protectorate, has had an interest-free paper currency since 1816, has no public debt, no unemployment, and a high standard of living.

Tax reform is also essential. "Geonomic" tax theorists, who trace their lineage to 19th-century American economist Henry George, argue that society should tax land and other basic resources—the birthright of all—instead of income from labor. Geonomic tax reform, say advocates, could decrease wealth disparities while reducing pollution and discouraging land speculation. Similarly, taxing nonrenewable resources and pollution—instead of giving oil companies huge subsidies in the form of "depletion allowances"—would put the brakes on resource extraction while giving society the means with which to fund the development of renewables.

Population and immigration. Overpopulation is currently one of humanity's greatest problems, and it will become a far greater one with the gradual disappearance of fossil energy resources. But of all the conundrums that beset our species, overpopulation is the most difficult to address from a political standpoint. Both the Left and the Right tend to avoid the issue of continued population growth, or treat it as if it were a benefit. As Russell Hopfenberg and David Pimentel have written,

> [s]ome people believe that for humans to limit their numbers would infringe on their freedom to reproduce. This may be true, but a continued increase in human numbers will infringe on our freedoms from malnutrition, hunger, disease, poverty, and pollution, and on our freedom to enjoy nature and a quality environment.

Currently, the regulation of population is probably best dealt with at a national—as opposed to an individual, community, or global—level because only nations have the ability to offer the incentives and impose the restrictions that will be necessary in order to reverse population growth. The first order of business will be for each nation to gain some sense of its human carrying capacity. Quite simply, if in order to maintain itself a nation is drawing down either nonrenewable resources (such as fossil fuels) or renewables at a faster rate than that at which they can be replaced, then that nation is already over-populated. A cursory scanning of population/resource data would suggest that virtually every nation on Earth has overshot its carrying capacity. This being the case, what should be the target size of national populations? The answer obviously varies from country to country. Globally, according to Hopfenberg and Pimentel,

> [i]f all people are to be fed adequately and equitably, we must have a gradual transition to a global population of 2 billion. A population policy ensuring that each couple produces an average of only 1.5 children would be necessary. If this were implemented, more than 100 years would be required to make the adjustment.

However, this global target needs to be translated into national goals and policies. Again, this is no small challenge.

A frequent tactic in this regard is to appeal to "demographic transition" as an ultimate solution to population problems. In the wealthiest countries, population growth has tended to slow. Germany and Italy, for example, currently have birth rates that are slightly lower than their death rates, which means that their populations are beginning to shrink. This suggests a painless solution to population problems: simply increase economic growth in other countries so that they undergo a similar demographic transition to zero or negative population growth. However, as should be clear by now, this is not a realistic option. Industrial growth cannot be maintained much longer even in Europe or North America; much less can we envision fully industrializing all of Africa, South America, and Asia. Another approach must be found.

The empowerment of women within societies also seems to result in reduced population growth. It is women, after all, who give birth and who traditionally provide primary care for young children; given the choice, most women would prefer to bear only a few children and see them grow up healthy and well-fed rather than have many children living in deprivation. Experience also shows that the ready availability of birth control methods and devices is, for obvious reasons, a significant factor in reducing population growth. These strategies if expanded, will certainly help; however, they probably cannot be counted on to produce the reductions of population size that are actually required in order to avoid famine and public health crises in the coming century.

During the past two decades, China engaged in a unique experiment at population reduction, attempting to limit families to having only one child. Describing this experiment, Garret Hardin writes:

> In some of the major cities the program seemed to be carried forward along the following lines. Decision making was decentralized. Almost every able-bodied woman in a Chinese city was a member of a "production group," which was charged with making its own decisions. Each group was told by the central government what their allotment of rice would be for the year. This allotment would not be readjusted in accordance with the Marxist ideal of "to each according to their need." Rather, it was a flat allotment that made no allowance for increased fertility. It was up to the members of a production group to decide among themselves which women would be allowed to become pregnant during the coming year.

If a member of a production group became pregnant without having obtained permission to do so, she was told to have an abortion.

The results of the Chinese experiment remain unclear since reports reaching the West have been vague and incomplete. There was no doubt a great deal of cheating involved, and farmers and many tribal groups were systematically exempted from the program.

A secondary effect of the Chinese effort occurred in the US, where reports of compulsory abortions in China incited rightist politicians to deny aid to Planned Parenthood and other organizations working to reduce population growth. Many international population programs were consequently seriously undermined by the withdrawal of US participation.

Clearly, the Chinese model—even if it can be said to have been successful in China itself, which is doubtful—will not work everywhere. What other methods are possible?

Ecologist Raymond B. Cowles once suggested using economic motivations to reduce fertility. He proposed simply paying young women *not* to have babies. The expense of such a system would be offset by the savings to society from costs not incurred for education and health care for the children who would otherwise have been born. Economist Kenneth Boulding proposed a somewhat similar *laissez-faire* solution: Instead of money, women should be granted, at birth, a certain number of "baby rights," which could be sold or traded. Lovers of children could buy such rights whereas lovers of money would be encouraged to devote their efforts to activities other than parenthood. Neither Cowles's nor Boulding's idea has garnered much support so far, but both illustrate the kind of creative thinking that must occur if we are to tackle the problem of overpopulation.

Immigration is an issue closely related to that of overpopulation, and it is likewise politically prickly. In the US, roughly 90 percent of the projected population growth for the next 50 years will come from immigration, with the national population projected to double during that time. Such population growth threatens to dramatically increase resource depletion and pollution. From an ecological perspective, immigration is almost never a good idea. Mass immigration simply globalizes the problem of overpopulation. Moreover, it is typically only when people have become indigenous to a particular place after many generations that they develop an appreciation of resource limits.

Opposition to uncontrolled immigration is often confused with anti-immigrant xenophobia. Also, some leftists cogently argue that to cut off immigration to the US from Mexico and other Latin American countries would be unfair: immigrants are only following their resources and wealth northward to the imperial hub that is systematically extracting them. Thus key elements in immigration reduction must be a halt to the US practice of draining wealth and resources from nations to the south, as well as democratization and land reform in the less-consuming countries.

In the decades ahead, all nations must find practical, humane solutions to the problem of population growth and immigration—solutions that will necessarily include legal caps on yearly immigration quotas and some means for reducing both disparities of wealth between nations and the exploitation of one nation by another so that immigration becomes a less attractive option.

As Virginia Abernethy of the Carrying Capacity Network has put it,

> [o]ften, allowing ourselves to be ruled by good-hearted but wrong-headed humanitarian impulses, we encourage ecologically disastrous responses among ourselves and our less fortunate neighbors. Impulses, which seem in the short run to do good, but which lead ultimately to worldwide disaster-and most quickly to disaster in the countries we wish to help-are not in fact humanitarian.

US foreign policy. America's military and espionage budgets represent a gargantuan investment in an eventual Armageddon. The US portrays itself as the global cop keeping order in an otherwise chaotic and dangerous world, but in reality America uses its military might primarily to maintain dominance over the world's resources.

This policy is unjust, futile, and dangerous. It is unjust because people in many nations are denied the benefits of their own natural assets. It is futile because the resources in question are limited in extent and their exploitation cannot continue indefinitely and because, by becoming ever more dependent on them, Americans are ensuring their own eventual economic demise. And it is dangerous because it sets an example of violent competition for diminishing resources—an example that other nations are likely to follow, thus leading the whole world into a maelstrom of escalating violence as populations grow and resources become more scarce.

The US policy of maintaining resource dominance is not new. Shortly after World War II, a brutally frank State Department Policy Planning Study authored by George F. Kennan, the American Ambassador to Moscow, noted:

We have 50 percent of the world's wealth, but only 6.3 percent of its population. In this situation, our real job in the coming period is to devise a pattern of relationships which permit us to maintain this position of disparity. To do so, we have to dispense with all sentimentality … we should cease thinking about human rights, the raising of living standards and democratisation.

The history of the past five decades would suggest that Kennan's advice was heeded. Today the average US citizen uses five times as much energy as the world average. Even citizens of nations that export oil—such as Venezuela and Iran—use only a small fraction of the energy US citizens use per capita.

The Carter Doctrine, declared in 1980, made it plain that US military might would be applied to the project of dominating the world soil wealth: henceforth, any hostile effort to impede the flow of Persian Gulf oil would be regarded as an "assault on the vital interests of the United States and would be "repelled by any means necessary, including military force."

In the past 60 years, the US military and intelligence services have grown to become bureaucracies of unrivaled scope, power, and durability. While the US has not declared war on any nation since 1945, it has nevertheless bombed or invaded a total of 19 countries and stationed troops, or engaged in direct or indirect military action, in dozens of others. During the Cold War, the US military apparatus grew exponentially, ostensibly in response to the threat posed by an arch-rival: the Soviet Union. But after the end of the Cold War the American military and intelligence establishments did not shrink in scale to any appreciable degree. Rather, their implicit agenda—the protection of global resource interests—emerged as the semi-explicit justification for their continued existence.

With resource hegemony came challenges from nations or sub-national groups opposing that hegemony. But the immensity of US military might ensured that such challenges would be overwhelmingly asymmetrical. US strategists labeled such challenges "terrorism"—a term with a definition malleable enough to be applicable to any threat from any potential enemy, foreign or domestic, while never referring to any violent action on the part of the US, its agents, or its allies.

This policy puts the US on a collision course with the rest of the world. If all-out competition is pursued with the available weapons of awesome power, the result could be the destruction not just of industrial civilization, but of humanity and most of the biosphere.

The alternative is to foster some means of international resource cooperation, but this would require a fundamental change of course for US foreign policy. Daunting though the task may be, it is time to recast US foreign policy from the inside out and from the ground up so that it is based not on resource dominance but on global security through fair and democratic governance structures.

Such a policy shift would necessarily imply both a voluntary relinquishment of US claims on resources and a dramatic scaling back of the US military apparatus. The latter could be accomplished through a fairly swift process of budget cuts, whereby funds formerly devoted to the military would be earmarked instead partly for the dismantling of weapons systems and partly for the redesign of the national energy and transportation infrastructure.

This would have domestic repercussions. Lacking a basis in militarily enforced resource dominance, the US economy would shrink. But this must be seen in perspective: it is an inevitable outcome in any case. The US, as the center of the global industrial empire, does not have the choice of *whether* to decline; it can, however, choose *how* to decline—whether gracefully and peacefully setting a helpful example for the rest of the world, or petulantly and violently, drawing other nations with it into an accelerating whirlwind of destruction.

Such a unilateral US relinquishment of global dominance would, it could be argued, open the way for another nation—perhaps China—to take center stage. Might Americans wake up one day to find themselves subjects of some alien empire? It may help to remember that the inexorable physics of the *energy transition* preclude such an occurrence. In the decades ahead, *no* nation *will be* able to afford to subdue and rule a large, geographically isolated country like the US. Only small, weak, resource-rich nations will be likely targets for conquest.

Transportation. Because of their extreme dependence on car and truck transportation, the US and Canada are, relative to many other industrial societies, at a disadvantage. In the US, the Interstate Highway system represents a vast subsidy to the automobile and trucking industries. Since that system's inception in the 1950s, train transport has languished, with Congress continually reducing the already-small subsidies available for rail transport. In the half-century from 1921 to 1971 (the year of Amtrak's creation), Federal subsidies for highways totaled $71 *billion*; for railroads, $65 *million*. Rail transport has received a total of $30 billion over the past 30 years, whereas federal subsidies for highways in 2002 alone amounted to aviation and airports, $14 billion. In the same year, a mere $521 million were set aside for Amtrak. According to a study by the *International Railway Journal*, at $1.64, the US ranks between Bolivia and Turkey in mainline rail road spending per capita. Switzerland spends the most ($228.29) and the Philippines the least ($29). Urban light-rail systems in the US have fared little better.

America's decades-long shift from rails to highways has been justified by the argument that railroads work better in areas of high population density while highways are more practical where cities and towns are far-flung. Most of the US has a much lower population density than Western Europe and Japan, where rail services move people cheaply and efficiently. In the US, where the distances traveled are typically greater, airlines are more attractive for interurban travel. This argument makes some sense—but only as long as fuel is cheap.

Even with the development of higher-efficiency and alternative-fuel cars and trucks, the energy transition will not permit the continued operation of a national auto/truck fleet of the current size. Moreover, commercial air travel may soon be a thing of the past as jet fuel becomes more scarce and costly. Trains—while still running on fossil fuels—have, when well utilized, lower energy costs per passenger-mile than either cars or planes. Thus, for the US, one sensible course of action would be to immediately cease subsidizing highways and airlines and to begin investing in rails.

At the same time, auto companies would be well advised to put in high gear their research into smaller, lighter, more energy-efficient electric, hybrid, and even human-powered vehicles.

Ultimately, for people in industrial societies, the future holds less travel in store, regardless of the means of transport chosen. Economic survival will thus require reducing the *need* for transportation by moving producers, workers, and consumers closer together.

Activism. In order to feel that the sacrifices they are making during the energy downturn are fair, the people of any nation must be empowered to participate in the process of making decisions about how those sacrifices are allocated. However, the fundamental changes to national economies and infrastructures described above are not likely to be implemented through conventional political means—by citizens voting for candidates—because it is in the interests of most politicians to lie rather than to convey bad news. More people will tend to vote for the candidate who promises the rosier future, even if those promises are patently unrealistic.

Therefore the radical shifts needed can probably only happen as a result of the dramatically increased involvement of an informed citizenry at every level of a revitalized political process.

Unfortunately, the citizenry is currently neither informed nor involved, and the system resists fundamental change at all levels. Immense sums are invested annually to distract the public from substantive issues and to turn their attention instead toward consumption and complacency.

The small minority who are aware of the difficult choices facing society need to take heart and redouble their efforts to educate others, including government officials.

Activists could play a crucial role in the upcoming energy transition, as they have played in most of the important social advances of the past few decades. Activist-led social movements have helped end colonialism and the worst manifestations of racism, gained rights for women, and helped protect numerous species and sites of biodiversity. Today many activists are advocating a rapid transition to renewable energy sources, conservation, and the equitable distribution of resources. Moreover, they are leading the way in modeling nonviolent social change.

An example of the latter is Marshall Rosenberg, whose Center for Nonviolent Communication works internationally with such groups as educators, managers, military officers, prisoners, police and prison officials clergy, government officials, and individual families. Nonviolent Communication trainings evolved from Rosenberg's quest, during the civil rights movement in the 1960s, to find a way of rapidly disseminating much-needed peacemaking skills. Today he is active in war-torn areas (Israel, Palestine, Bosnia, Columbia, Rwanda, Burundi, Nigeria, Serbia, and Croatia), promoting reconciliation and a peaceful resolution of differences. As social and economic pressures from the energy transition mount, such mediation efforts could—both globally and locally—mean the difference between peaceful cooperation and savage competition.

Social activists tend to be the leading edge thinkers and change agents for society as a whole. We need more of them.

The World

Many people are wary of world government, believing that it would lead inevitably to global tyranny. This fear is both founded and unfounded. It is well-founded in the sense that people's individuals ability to contribute to the decisions that affect their lives varies inversely with the scale of social organization: it is easier to make one's voice heard in a town meeting than in a national election. This being the case, it seems highly likely that a world government, were one ever to be established, would tend to be remote and unresponsive to the needs of individuals and local communities. But the fear is unfounded in the sense that, without fossil fuels, it is doubtful that a sufficient energy basis could ever be assembled to build and maintain a government with a global scale of organization, communication, and enforcement.

Hence the reasonableness of a principle succinctly stated by ecologist Garrett Hardin: *Never globalize a problem if it can possibly be dealt with locally.*

Are there any problems that must be dealt with globally? In ordinary times, there probably are not. However, during the extraordinary period of the peaking and decline of fuel-based industrialism in which we are now living, there are three kinds' of problems that do indeed demand some kind of global regulatory mechanism: resource conservation, large-scale pollution control, and the resolution of conflicts between nations. All three must be *administered* more or less locally: Resources exist in geographically circumscribed areas that are ultimately the responsibility of regional decision-making bodies; pollution often issues from point sources that are best monitored by local agencies; and conflicts must ultimately be resolved by the parties involved. But the depletion of internationally traded essential resources, industrial production processes, and industrialized warfare are capable of having overwhelming global effects. Catastrophic global warming and nuclear war provide compelling examples: either would result from decisions made, and actions taken, by specific people in particular places; but the consequences of those decisions and actions would profoundly impact people and other organisms everywhere. The consequences are so far out of proportion to the decisions and actions taken locally that some form of global control mechanism seems called for, consisting of enforceable minimum conservation standards and enforceable means of containing or resolving international conflicts.

Some agencies already exist for addressing global problems. They are generally of three kinds: first, corporations, trade bodies, and lending institutions; second, the quasi-governmental apparatus of the UN, with its related aid agencies; and third, the small but vocal cadre of transnational human rights and environmental NGOs.

The corporations, international banks, and trade bodies together constitute a force for globalization-from-above. They are doing almost nothing to help, and much to hinder, an orderly global energy transition. This should be no surprise: they are part and parcel of the growth economy that flows from the fossil-fuel pipeline.

The forces of globalization-from-below (the NGOs) do not have a full picture of the degree to which world events revolve around energy resources and their depletion; nor do they have an adequate strategy for dealing with the issues they are confronting. But their push toward decentralization, democratization, and cooperation is nevertheless generally the right way to help humanity wean itself as painlessly as possible from fuel-fed industrialism. Thus what is needed globally is a weakening of the forces of globalization-from-above and a strengthening of those of globalization-from-below.

The UN—which is caught somewhere between those two sets of forces—is one of the few institutions that is in any position to provide enforceable minimum global environmental standards and to serve as an arena for conflict resolution.

If all parties concerned understood the severity of the crisis facing them, there is much they could do. They could negotiate more global agreements modeled on the Kyoto accords, ensuring international efforts to reduce greenhouse gas emissions and subsidize renewables. The International Energy Agency could be expanded and empowered to survey, conserve, and allocate energy resources in such a way that all nations would have assured (though diminishing) access to them, and that profits from resource exploitation would go toward helping societies with the transition, rather than merely further enriching corporate executives. Meanwhile, UN-based conflict-resolution and weapons-destruction programs could substantially reduce the likelihood of violent conflicts erupting over resource disputes. Rich industrialized nations could wean themselves as quickly as possible from fossil fuels while less-industrialized nations, abandoning the futile effort to industrialize, could embark on the path of truly sustainable development. Industrialized nations could assist the latter in doing so by ceasing the practice of siphoning off less-consuming nations' resources.

What is especially needed is a new global protocol by which oil-importing nations would agree to diminish their imports at the rate of world depletion—approximately two percent per year. That way, price swings would be moderated as the peak of global oil production passes, enabling poorer nations to be able to continue importing the bare minimum of resources needed to maintain their economies. A Model for such an agreement has been proposed by the Association for the Study of Peak Oil (ASPO).

The majority of the world's nations and peoples would probably be willing to participate in all of these difficult and even painful efforts if they were informed clearly of the alternatives. The greatest impediment would likely be the non-participation of a few "rogue states" that tend to disregard international laws and treaties at will. The foremost of these are the US and, to a lesser extent, China.

With only five percent of the world's population, the US has the lion's share of the world's weaponry and exercises direct or indirect control over a steeply disproportionate share of global resources. The US cleans up some pollution at home while undermining international environmental agreements. It refuses international inspection of its weapons of mass destruction and attacks other nations virtually at will. It also undermines efforts to stabilize or reduce the global population at every turn. Just within the past four years, the US has abrogated the anti-ballistic missile treaty and undermined the small arms treaty, the UN convention against torture, the international criminal court, and the biological weapons convention.

Will the US join the international community, or insist on maintaining its privileged status even as its empire crumbles? This is the first great geopolitical question we face as the industrial interval wanes.

The second one concerns China. Will China continue to seek to industrialize? Because of its huge population, efforts in that direction will put great stress on any global efforts at conservation and pollution abatement.

The world does not revolve around these two countries. But if they could be persuaded—by either their own citizenry or the international community—to exercise constructive leadership, the global energy transition could occur far more smoothly than would otherwise be the case.

Taken together, these recommendations imply a nearly complete redesign of the human project. They describe a fundamental change of direction—from the larger, faster, and more centralized to the smaller, slower, and more locally based; from competition to cooperation; and from boundless growth to self-limitation.

If such recommendations were taken seriously, they could lead to a world a century from now with fewer people using less energy per capita, all of it from renewable sources, while enjoying a quality of life that the typical industrial urbanite of today would perhaps envy. Human inventiveness could be put to the task of discovering ways not to use more resources, but to expand artistic satisfaction, find just and convivial social arrangements, and deepen the spiritual experience of being human. Living in smaller communities, people would enjoy having more control over their lives. Traveling less, they would have more of a sense of place and of rootedness, and more of a feeling of being at home in the natural world. Renewable energy sources would provide some conveniences, but not nearly on the scale of fossil-fueled industrialism.

This will not, however, be an automatic outcome of the energy decline. Such a happy result can only come about through considerable conscious effort. It is easy to imagine less desirable scenarios.

Save Three Lives

A Program for Famine Prevention

By Robert Rodale

This was probably the hardest chapter intro of all for me to write, because if I could I would include Robert Rodale's entire book. His style is chatty and engaging, but it would be a pity if that made you take what he has to say less seriously. Rodale, Inc. is the grandfather of organic agriculture in the United States. My maternal grandfather, a Virginia country lawyer, used to read Rodale books, buy Rodale products and grow his precious tomatoes and okra by Rodale principles. Robert Rodale (son of the founder of Rodale, Inc.) tackles the difficult subject of famine and makes very difficult points in this book.

Our readings begin with a description of how colonial policies made Africa prey to famine by encouraging corn farming, rapid population growth and deforestation. The next chapter, titled "How Foreign Aid has Only Increased the Vulnerability of Hungry People," demonstrates that foreign aid typically goes to foreigners, for projects that benefit themselves, not local people. Such projects often move locals from subsistence agriculture into the production of cash crops, with the idea of bringing them into the global economy. Such projects tend to concentrate the money and power they generate in the hands of the few, while removing fertile land from the production of food for local people. In addition, these projects typically focus on *monoculture*, growing one crop only on land where peasants have traditionally intercropped many species to create robust, interdependent agro-ecosystems.

Guatemala is a good example (mine, not Rodale's). You can grow good coffee in Guatemala, so aid projects favored turning peasant farms into coffee plantations. Now, when coffee prices crash, peasants go hungry. Actually they go hungry all the time, since (with the help of the 1954 CIA-sponsored coup) a few hundred families continue to control virtually all of Guatemala's wealth. Similarly in Africa: "The invasion of a large group of experts with their big plans for big dams and big plantations often changes the local balance of power that's been in place for centuries." From this chapter, I have included Rodale's description of the Green Revolution, the mother of all such projects.

Rodale goes on to extol the virtues of a small, NGO-based, microcredit approach to rural empowerment. This is similar to the "Grameen Bank" approach: A fund is set up to make small loans for peasants to improve their farms. Tremendous importance is placed on repaying the loans, seen as a matter of village honor as well as practical necessity. To quote one of the speakers in the remarkable film *FLOW (For Love of Water)*, "The World Bank knows how to loan a million dollars in a thousand places. What it doesn't know how to do is loan a thousand dollars in a million places."

Save Three Lives primarily concerns Africa, though most of its conclusions apply to the rest of the world as well. Africa is often in the headlines for famine; Somalia is the latest case. Yet Chapter Four of this book is called "Famine Relief is Not the Answer." This provocative title underscores a tragic fact. As Rodale puts it, *"the relief efforts to which we give our money simply don't work.* [...] *you have to understand that you cannot help the people in those pictures. Many are already dead by the time you see their image. And most of the rest are too far along the path of malnutrition to come back to anything but a painful, crippled life—even if your donated food could reach them in time."* Of the Ethiopian famine of the early 1980s, he notes that "Much of the food arrived, ironically, 'as local famers were bringing in their first good harvests in years'. Harvests that, unfortunately, were now worthless to those impoverished farmers because the country was flooded with free grain."

Rodale is not heartless; he simply asks that we turn our concern back in time, back from the dramatic images of starving people on the march to the image of the struggling farmer. Before they march, before they starve, before they leave the land and migrate to the city, while they still cultivate and care for the precious land; that is the time to help them, and that is why *Save Three Lives*, a book about helping peasants, is subtitled *A Plan for Famine Prevention*.

The central portions of the book, therefore, concern themselves with methods for helping farmers maintain soil fertility, grow sufficient food, and have wood for construction, fencing and fires. Trees are central to his approach, which is a type of *agroforestry*. Rodale favors an approach called "alley cropping" in which rows of trees are planted in between rows of crops. The trees pull nutrients up from deep in the soil, below where the roots of food crops can reach. The leaves and branches of these trees can be used as fertile mulch, providing nutrients for the food crops, building soil, and retaining the rainfall as soil moisture. Obviously, the trees provide wood as well.

There are numerous subtle and excellent points made in these chapters. Most of them are beyond the scope of this class, but I will mention some tantalizing ideas that underscore how little we really know about feeding the world, and the extent to which these ideas make industrial agriculture look about as appropriate as cutting butter with a chainsaw:

- Amaranth and other wild plants[1] are gathered and eaten during droughts as "hungry food." These have enormous potential to be bred into crops that are both more nutritious and use less water than corn and our other current staples.

---

1   If you want to follow this up, you could start with the African yam bean, Bambara groundnut, desert date, Hausa potato, "hungry rice," Marama bean, tamarind, teff and ye-eb. Read all of *Save Three Lives* to find others!

- Cultivating such crops will restore cultural pride to peoples whose agricultural legacy has been cast aside and devalued in the rush to industrialize agriculture.
- Low water use crops could greatly reduce the need for ecologically damaging irrigation projects.
- Native plants will combat the desertification that follows upon deforestation, monocrop agriculture and subsequent soil depletion.
- Some tree species, in particular, not only put down water-seeking roots as deep as 100 feet, they can also absorb fog and night mists. Their roots hold soil, preventing it from blowing away, and preventing sand dunes from forming. Reforestation thus reverses desertification.
- Alley cropping with leguminous (nitrogen-fixing) trees will add nitrogen, a key plant nutrient, to the soil without the need for expensive and ecologically damaging nitrogen fertilizer.
- Tree leaves make excellent animal feed. Animals provide milk, eggs and meat for food, manure for fertilizer, and energy to plow fields.
- The leaves of certain trees have huge potential to provide nutritious food for humans as well. Nigerian tribes eat the leaves of 22 different trees. This is but one example of the agricultural and famine-prevention legacy that is being cast aside when we ignore traditional peoples.
- Alley cropping and other agroforestry practices create an agro-ecosystem that is much closer to the original wild ecosystem, encouraging native biodiversity to flourish. This is vital given the enormous medicinal, food and other values of wild plant species both known and unknown.
- The mulch and shade provided by agroforestry makes toxic and expensive herbicides unnecessary.
- Agroforestry encourages beneficial insect species that eat crop pests. Most Americans are familiar with ladybugs and praying mantises. *Trichogramma* wasps are another example. These "biological controls" eat the bugs that eat our crops.
- The modern industrial model attempts to kill *all* crop pests. The agroforestry model kills *most of them.* It turns out that leaving a few is actually beneficial! "Over generations, they may develop new characteristics that make them more resistant to attack by these pests. In the short run, however, and insect attack simply signals the plants to grow stronger and faster. ... If there were no attack, the plants... would get lazy. Their natural defenses wouldn't kick in."
- Some plants actively repel insects, and these can be cultivated alongside food crops.
- Many traditional food plants have been bred over thousands of years to be resistant to crop pests. These varietals should be cultivated instead of Green Revolution hybrids.
- Plants such as vetiver grass can be used to retain soil in hilly areas without expensive land contouring.
- Improved composting, especially via composting pits, is a simple technique that can be taught to villagers to improve soil.
- Simple water harvesting techniques, such as leading rainfall runoff to planted areas, can make desert farming possible. *Save Three Lives* describes the related techniques of catchment harvesting, water diversion and spreading, microcatchment, contour catchment and artificial springs.
- Aquaculture (fish farming) can be a valuable source of protein.
- Where brackish water is a problem, salt-tolerant food plants can be cultivated.

My friend Malaki grew up embedded in a culture that is one of Africa's richest, in a Luo village in Kenya, using cattle to till the fertile soil. Over the years he has watched his village wither as young men, in particular, go off to the city to seek their fortune. He says that such shame would be associated with failure that no one returns—or if anyone does, he shows up in a cheap suit, lies about what a big man he has become, and leaves as soon as possible. Malaki has started an NGO called Grow Strong here in San Diego that aims to help villages maintain themselves, and to give these tragically displaced young men the permission and the emotional space to return to their traditional lives with pride. There are many such small NGOs, and to the extent they are successful, not only will villages be aided, but population pressures will be greatly eased in urban squatter colonies and developing world cities in general.

In the meantime, however, throughout rural Africa, women do the farming, in part because so many of the men have gone to cities. Some men simply disappear; others, if they are modestly successful, send home a small remittance from their urban jobs. But the women continue to bear and raise the children, do the terribly difficult farm labor, and suffer from the above-mentioned, misguided rural development efforts. That is why, of all the pages I wanted to include from *Save Three Lives*, I chose to close with sections from a chapter on how empowering rural women will fight famine and ease population pressures.

As people urbanize, birthrates drop and populations stabilize after a generation or so, for a number of reasons too complicated to discuss here. This is called the "Demographic Transition" and is a big reason why human population is expected to stabilize at around nine billion in the middle of this century. Rural areas tend to have the highest birthrates, driving rural-to-urban migration and continued human suffering. Rural women are the most disempowered population in the world; given a choice, they usually seek an education, learn about contraception and have fewer children. I won't give away too much of this chapter; I'll just say that Rodale makes some startling statements about the plight and the potential of rural women.

Africa is not actually overpopulated. It has enormous resources and a low population density compared to, say, Asia or Europe, which are in clear ecological overshoot. Africa's problem is its colonial legacy. Most of its resources are being shipped away to richer parts of the world, to the benefit of foreigners and corrupt local elites. Rodale pleads eloquently that appropriate-scale, ecologically sensitive modern techniques can help revive village values that sustained this suffering African continent for many generations before it was plundered.

How the Lushest Land in the World Fell Prey to Famine

[Old African hands point] to areas of desert and barren hillsides, and tell you of lush grasslands and fertile mountains that used to be. Of majestic wild animals that used to roam and hunt, enriching the land beyond description with their grace and beauty.

I met one of those men in Ethiopia, in 1985, the year after the last really big famine there. His name is John McMillan, and he was working for the organization *World Vision* at the time. In 1985 he was about sixty-five years old, and had been in Ethiopia forty-five years ago. He went there as a teenager right after World War II. The very area where we met was a place in which he had spent a considerable amount of time four decades previously. He pointed to a treeless hillside.

"When I was here in the '40s that hillside was covered with trees," he told me. "And there were lions," he said with the sadness of a man who may have been a bit afraid of the great cats then, but missed their roars and majesty now. Listening to his voice, you could imagine a female lion bringing down a gazelle in the tall grass while the green mountains shimmered in the distance. You could see an abundance of life.

"It was the breadbasket of Africa," he said. Now, remember, he's talking about Ethiopia. A name much more likely to bring to mind images of starving children and barren wastelands—not lush mountains and lions lounging in the sun.

"Mussolini marched in here for a reason," he reminded me. Like the Roman conquerors centuries before, the Axis powers saw the tremendous agricultural potential of Africa. They saw a farm the size of a continent that could feed the millions of soldiers fighting their awful war.

"And that lushness was typical of Africa at the time. It wasn't just this area," my friend explained, "Africa was full of trees and all kinds of animals.

"All gone now," he added, with a sad shake of his head. "All gone now.

So what happened? Even though it's a mistake to try to talk about "Africa" as if that immense continent were one simple place (just the area consistently affected by drought is three times the size of the United States), you can safely say that as recently as a hundred years ago traditional methods of farming were the norm.

Because Africa is so large and so radically different from place to place, those traditional methods took many forms. The kind of system that worked well on a lush hillside or near a river delta would be very different from farming that was successful in an area where local water is virtually nonexistent and rains are scarce, for example.

But no matter where they were practiced, those traditional forms of farming did have several things in common. One is that the people farming these different types of land all grew a mixture—a wide variety—of crops. It would have been unthinkable to simply plant massive amounts of one crop (not to mention nutritionally dangerous, and pretty boring at mealtime). Everyone grew variety—a mixture of grains, tubers, and fruits, as well as green plants and other crops that sometimes supplied food and at other times supplied the ingredients for local medicines.

Another common element is that most African farmers also practiced a system of rotation. After a piece of land had been used for a few years, it would be pretty well drained of its natural nutrients (this is especially true in the tropics, where the soil is much less rich than you might think). Typically, the farmers either burned the fields or simply moved to another area for a while. Or both. Years later, after the land had rested and regenerated itself, they or someone else would return and farm that space again. Basically, their natural instincts led them to develop a fallow system—much like the government still pays some American farmers to follow—that protected the land and ensured good harvests.

Then the Europeans arrived. We know all too well that initially they farmed the people themselves—abducting the African natives and selling into slavery those that survived the hellish journey across the ocean. But it didn't end there. Unfortunately, history doesn't pay as much attention to the second incursion of the Europeans. After the human slave trade dried up, the Europeans enslaved the land itself.

They saw Africa as a place to be conquered, colonized, and managed. They saw tremendous areas of lush land—land whose potential, they felt, was being wasted by primitive people with primitive cultures and (to European eyes) primitive methods of agriculture. Colonized, Africa's land created tremendous wealth for the Europeans. But the manner in which they used that land set in motion a chain of events that has contributed to the problems the continent faces today.

For one thing, the Europeans brought corn to Africa. And, unfortunately, corn grew very well there. I say, "unfortunately," because Africa today would be much better off if the land hadn't been so well suited to corn. Maybe the Europeans would have given up and gone off and abused someone else's land. Maybe they would have used Africa's land in a less destructive manner. But the corn grew well, and so they stayed. And the massive amounts of corn they grew—again, we're talking about a lusher, much more productive Africa back then—had two basic effects that helped lead to the problems of today.

Too much corn leads to too many people. Here in America we also grow a lot of corn. We eat some of that corn, but a huge proportion of the harvest goes to feed animals.

That's not the case in Africa. Throughout most of the central part of the continent, you find the tsetse fly, which likes to bite cattle on the back of the neck. The cattle get sleeping sickness and die. Oddly enough, people can manage there better than cattle. And so, there have always been vast amounts of land, much of which was extremely fertile, which could be farmed, but couldn't be used for livestock, or even to grow food for livestock.

So corn was grown. And people ate the corn. Over the years African people have become accustomed to growing and eating a type of corn that is very different from anything that most Americans have ever seen. It's a white corn, not yellow. But it's not the kind of supersweet white corn—like Silver Queen—that we eat in the summer. It's a white field corn that is not picked fresh, but left out on the stalk until it dries and hardens. After picking, the people grind it up into cornmeal and eat it in a number of ways.

It is delicious, by the way. I love to eat this corn. In fact, I try to keep a steady supply around, and I make all my corn bread from white field corn. Like Africans, I just choose not to eat yellow corn-meal. The yellow is considered more nutritious, but the white has a powerful, wonderful flavor in the mouth.

So a tremendous amount of corn was—and still is—grown. But most of the corn was not grown for export. The European colonizers—and, more recently, the development experts—had long-range, far-reaching plans. They knew that Africa could produce a huge amount of coffee, cocoa, sisal, citrus, palm tree products like coconuts and palm oil, and expensive nuts like cashews. Those were products for which there was a good cash market in the West. And they also required a lot of cheap labor to produce.

So the colonizers set out to "fatten up" the people. They fed them lots of corn and introduced the concept of Western-style public health so that there would be lots of locals to work the vast plantations of cash crops. Of course, some of the public health improvements, especially the more recent ones, were done for humanitarian reasons as well as fiscal, but the result was the same: Because of the abundance of corn and the improvements in health care, the population soared.

Now, obviously, no one is going to say that health care improvements such as better sanitation and infant vaccination shouldn't have been introduced to Africa. In fact, the increase in population caused by those improvements in public health doesn't seem to have been the real problem.

Most experts feel that Africa's own natural resources could have handled the population increase that was the natural—but totally unforeseen—result of introducing modern medicine. *If* the change had been limited to medicine and public health improvements alone. But it wasn't. There was more food as well. More food and better medicine meant more people were being born, and more of those people were surviving. Which meant more food was necessary. And, of course, more food meant more people....

Initially, it didn't seem like a bad strategy. And when some people finally did become concerned that there were no longer any natural checks on the population level, their solution, unfortunately, was simply to grow more corn.

But the corn was doing several things to the land that would soon make that solution impossible. If you've ever grown corn, either on a farm or in your garden, you know what a hungry plant we're talking about. Corn devours nutrients. You have two choices with corn: Feed it vast amounts of fertilizer or plant in a different place every season. So the first, and simplest effect of all this corn was to rapidly suck the land dry. Corn is not user friendly—the plant has a huge appetite and doesn't give very much back to the soil after the harvest.

In fact, under most circumstances, corn isn't even cost effective. It's very difficult to get enough money back from the ear or two that a plant produces to justify the fertilizer that you put into growing those ears. (That's why corn production almost everywhere has to be subsidized to some degree by governments. And even with those subsidies, a farmer will never get rich growing corn.)

In the development plans that were drawn up in the 1960s and 1970s to address the problem of out-of-control population growth in many Third World countries, massive amounts of fertilizer were shipped into Africa (and other developing nations) from overseas—at enormous cost, of course. This fertilizer—all chemical based, mostly derived from petroleum—didn't do anything to improve the soil, but it did keep the corn growing.

Before the fertilizer arrived, and in areas it never reached, the fields were burned and abandoned after each harvest. Again, with a manageable population, such a method might work. If you were lucky and conditions were right and local plants popped up

quickly to keep the soil in place, and you let the fields regenerate for a decade or so, you could probably grow another corn crop there.

But as the population grew, long fallow periods became impossible. People tried to use the land again much too quickly. And, of course, a farmer—anywhere—should never count on being lucky. Right after the harvest—especially if the field is burned—the land is in great danger. Because there are no longer any plants in the soil to keep that soil in place. All it takes is a hard rain or a strong wind and that soil is gone.

And the more land that's exposed at any one time, the worse the problem becomes. For unless a driving rain is broken down by plants—trees are the best choice—those seasonal torrents of tropical water are not absorbed by the soil. Instead, those rains wash away soil with incredible efficiency.

Soil not held tight in the ground by roots is also at the mercy of the wind. And Africa is home to the kind of merciless, legendary winds that they name fast cars after (the scirocco, for example).

No wonder the soil losses in Africa are so staggering. (As are the losses in America, by the way. Many areas in the United States can now only grow food by using enormous amounts of chemicals. There's simply no soil left—just enough little bits of pulverized rock and sand to hold the crop roots in place. And that's why famine is such a distinct possibility in America as well as in the Third World.)

Slowly, over the course of the last hundred years, huge tracts of ecologically sound land in Africa have been cleared, used for a couple of years, abandoned, and then claimed by the desert. Thousands of native plants and millions of acres of grasslands were lost. Millions of trees were cut down. All so that vast stands of corn and cash crops could be planted.

Those crops not only left the land helpless after their harvest, but also stripped the land of all the nutrients that the fragile tropical soil had built up over millions of years of natural cycles of death, decay, and then regeneration.

But because Africa was so lush when all this started, and because Africa is so vast, the process was able to go on for a century before being crushed under the weight of its own foolishness. Lots of corn was grown. The population exploded in response to this enormous increase in the food supply. Slowly, at the same time, the same corn was also causing the land to become less and less productive.

In retrospect, it's hard not to have seen famine coming. It was inevitable—just a matter of time.

But the real problem is wood. Africa was "Rome's woodlot." Wherever did it all go?

Some of those trees—most of them—were cleared for agriculture, with the result that we just discussed. Sadly, no lesson has been learned, it seems. In other parts of the world, right now forests worth billions are being cut and torched in order that cattle can graze (for a surprisingly short period of time, it turns out) so that we can eat cheap hamburgers at fast food restaurants. Unimaginable numbers of fabulous, spectacular forms of life of all kinds are being wiped out. Animals, reptiles, plants, and microscopic creatures found nowhere else in the world are being made extinct before we can find them, study them, analyze them. Before anyone can examine them in the ongoing search for new medical discoveries based on nature; the cures we're all desperately hoping science will provide. But those discoveries—those cures—can never come because we're selling off the natural laboratory at flea market prices! Sorry about that endangered tree frog with the promising venom that might have reversed rheumatoid arthritis. Oops, looks like we lost this rare tropical orchid with that strange pigment that might have had the potential to treat cancer.

Here, have a burger instead.

How Foreign Aid Has Only Increased the Vulnerability of Hungry People

A bit of Green Revolution background: To combat hunger in India and Asia, vast areas were planted with a scientifically bred single strain of super-rice that (perhaps due to its artificial nature) had a huge appetite for chemical fertilizer and a stronger-than-average need for pesticide to protect it against hungry insects.

Once those chemicals were applied in large amounts, the immediate effects were truly astonishing. Yields responded dramatically. Record increases in harvests were recorded. But as the World Resources Institute's book, *To Feed the Earth,* wisely points out, "Record crops produced at the expense of next year's or the next decade's soil resources are nothing to be proud of."

One failure of the Green Revolution is that by relying exclusively on chemical fertilizers it added nothing real to the soil. The should-have-been-foreseen result? When the inevitable happened, when the programs funding these massive supplies of chemicals ended (or when the chemicals themselves became illegal or just too expensive to buy), the land was left in much worse shape than before the "revolution." Another inevitable long-term failure that should have been foreseen was the poisoning of water supplies by the huge amounts of chemicals necessary to grow artificially created crops under such unnatural conditions.

Perhaps these things *were* foreseen. Perhaps the end was felt to justify the means. Or perhaps the experts never thought to look toward the future, preferring instead to luxuriate in the comforting present of bountiful crops. What they failed to anticipate, however, was that their period of reward—their "window" of success—would be so brief. In terms of time, the chemically cradled process known as the Green Revolution has been more of a peephole than a window.

[Rodale then gives examples from Chiang Mai province in Thailand and from China. Initial, dramatic gains in grain yields from Green Revolution seeds, fertilizer and chemicals are undercut in subsequent years. Pollution becomes significant and productivity decreases as soil is damaged.]

[Also, monoculture farming makes] pests' life easier. The so-called development experts point to their endless seas of genetically engineered super-rice and say, "Just look at all that food!" Standing there next to them, I see those same endless fields of identical plants and think, "Just look at all that vulnerability!"

Yes, vulnerability. Vulnerability to one pest, to one disease. It's a given in agriculture that growing several types of plants close to each other confuses pests. Mixed plantings slow them down, preventing their spread by presenting them with natural borders.

But a single crop that goes on for miles? All you need is one enterprising type of insect or determined disease to catch on to this inviting situation and that crop killer could conceivably spread like a prairie fire, devastating everything in its path (a path conveniently built by human hands, no less). That's because, in such a foolish form of agriculture, everything is really just *one* thing. And that suits single-crop insects like corn borers just fine. Before plantation farming, these pests were present, but relatively under control. Typically, they would find a small stand of their preferred crop, settle in, and stop when they reached the edge of any part of that field.

Pests such as the borers, that are specific to corn, won't cross an acre of broccoli or beans in a healthy, mixed-planting system—even if there is corn on the other side of those less-tasty plants. So they stay where they are, content. Their population, with a limited food supply, remains under control.

Insects that eat a wide variety of plants, such as the army worm, don't like to cross open land. As long as open spaces surround the stand they occupy, they'll stay where their plants are. And, again, the limited amount of food controls their population naturally.

But give insects endless acres of corn or any other crop back-to-back and it's like building a superhighway, a free public transit system for pests. You couldn't make their spread any easier if you picked them up and hand-carried them from plant to plant!

[Rodale describes how a superweed called *Striga* wipes out whole fields of monoculture crops. Rather than using tried-and-true methods like planting multiple crops so that no one pest can completely destroy a farm, so-called development experts insist on breeding crops resistant to single pests. In 1977, in Indonesia, 2 million tons of rice bred to resist the brown planthopper were lost to a virus called *tungro*. This is not effective since the crops don't have universal resistance.]

And, of course, then it's simply a matter of time until another unexpected occurrence wipes out that variety as well (a hard piece of reality that, unfortunately, does not enter into the thinking of such planners).

But the appearance of an opportunistic insect or disease is not an unexpected occurrence to people who are paying attention, and it shouldn't be to the experts, either. Over several millennia, traditional farmers have learned that insects and disease are to be expected.

Both insects and disease are extremely efficient. Whenever we grow huge amounts of food for ourselves, it's fairly certain that something else out there is going to see that bounty as food as well. In a real sense, food creates life. And often that life is serious competition for us. If, by some miracle, we were even able to create some new kind of food that no insect currently eats, you can bet it wouldn't be long before one simply evolved to feast on that food. And pesticides, at least to those who have eyes to see, are not the answer to this simple reality of the food chain.

First, there are very few true pesticides, in the sense that they harm only pests. Almost all these poisons are extremely effective humanicides as well. The World Resources Institute points out that "as many as 400,000 illnesses and 10,000 deaths may be caused by pesticides every year worldwide—most of them in the developing world."

And second, although those chemicals are as much of a danger to us as they are to pests, the insects are remarkably better at adapting to those poisons than we are.

But the prevailing mentality is still "We'll dominate, we'll kill them. If the poisons that we've got don't work, well just make stronger poisons." And, of course, the highly adaptable insects become resistant to the new poisons as well, while the people who live in the area often wind up being sickened or killed by these increasingly toxic chemicals.

———————

[Rodale now describes how, because insects have such short reproductive cycles, they can rapidly evolve resistance to pesticides. When a field is dusted, a few insects survive, and they pass their accidental genetic resistance on to their offspring. He then takes up the topic of disease resistance, pointing out that "there is not here". The tropics have far more pathogens than temperate zones, so the very idea of breeding disease-resistant tropical crops is far more problematic than doing the same in a temperate climate like the United States; 54 diseases attack US rice harvests, but 600 attack tropical rice like the aforementioned Indonesian crop.]

———————

Tomatoes? In your backyard, this favored summertime crop may have to resist as many as 32 different diseases. Move to the tropics, put the same plants in the ground, and now you're up against 278.

These are not isolated examples. The difference exists across the board. The lesson to be learned is that there is not here. You can't farm the Sudan or the Philippines as if they were slightly displaced pieces of Iowa.

Forget for a moment the culture and the poverty—the fact that the people involved can't afford that U.S. style of high-priced farming (tractors, fuel, pesticides, herbicides). The differences are even more basic than that. Tropical soils are different, fragile almost beyond belief. Because there are no seasonal freezes, crop-eating insects are more plentiful. The diseases that threaten growing plants are almost unbelievably more abundant. Water is often scarce; sometimes it's practically nonexistent. And, in most areas, there simply are no roads to truck all that expensive Green Revolution equipment in. And where roads do exist, they would have to be improved greatly before we would consider them to be in merely terrible shape.

There is not here. And that's the single biggest reason why massive plans ostensibly designed to feed the people have often failed, taking people who were living on the edge of famine and pushing them right into its jaws.

You simply can't send U.S. farming equipment and know-how to another part of the world where nothing even vaguely resembles the United States and expect the system to work. Luckily, I see more and more people in official positions acknowledging that now. But this enlightenment has been a terribly long time coming.

Many years will be needed to reverse the damage, but eventually the hard-working, fertilizer-producing water buffalo will return in sufficient numbers to replace the rusting skeletons of the diesel-guzzling, pollution-belching tractors that have died early deaths due to a lack of spare parts.

A small, fearful aside: The failures of the Green Revolution in Asia and India are becoming more and more apparent every day. One can only hope that those in power have the courage to admit their mistakes before this unnatural form of agriculture begets a true, international disaster.

What might appear if this foolish form of chemical farming doesn't stop? A new insect? Perhaps a mutant locust, impervious to our most toxic chemicals, but with a sinister twist—say five times the breeding capacity or reproductive speed of its already deadly natural cousin? Or will it be a new disease? One that takes root in this unnatural landscape, but quickly spreads—perhaps on the wind—to menace the rest of the world ?

Previous famines may pale in comparison to the unknown horrors that could be unleashed.

Even though it's obvious that Green Revolution techniques could never work in Africa, people have tried to foster this bizarre concept in what was once the garden spot of the earth. Thank goodness, it never had a chance. Africa isn't Asia, and it isn't India. As Donald Curtis and the other authors of the British book *Preventing Famine* (1988) point out in their introduction, "The vast drought-prone areas of Africa are very different from the flat, irrigable or well-watered areas of the world where the 'green revolution' and other technologies have made their impact."

Again, even though Africa is the sum of many divergent parts, some things can be said about the continent as a whole. Africa is fragile. It also has great agricultural potential. The land fed vast numbers of people for millions of years before outside efforts simultaneously sabotaged its agricultural potential and artificially bolstered the population. Depending on how those in power now choose to treat its land, Africa could once again be green (a true and natural green, not the artificial green of the Green Revolution) or it could continue to turn brown. The color of the desert. Lifeless.

A vast area known collectively as sub-Saharan Africa is where those choices will have the biggest effect. It is here that people who make decisions about agriculture have a remarkable power to work for or against the desert. Lately, the people in power seem to have been virtually in the employ of the desert. The leaders have been very good to the sands, helping them reach out and spread their barren domain. Part of this sandy expansion has been achieved with the use of Green Revolution techniques: attempting to grow high-yield varieties of crops with heavy water requirements in arid lands. Again, anyone but a development expert could easily see the fault in such logic.

And yet the chronic lack of water—especially reliable water—is not the strongest point against the use of Western farming techniques in Africa. One of the strongest arguments against those techniques—and one of the strongest arguments for our own plan, our return to a very slightly modernized form of ancient agriculture—is Africa itself. The continent is simply not suited to the Green Revolution technique.

Consider the roads, for instance. Even if it made ecological sense to import those genetically engineered superplants and the chemicals and diesel-fueled equipment necessary to grow them, it simply can't be done, mostly because the roads just aren't there. Where there are roads, there are also soldiers, waiting to waylay such shipments (and relief convoys as well), kill the drivers, and steal the trucks and their contents. The chemicals, seeds, and equipment are then sold on the black market. Relief food goes to soldiers. Sometimes it is simply destroyed.

[Rodale goes into some detail about how bad roads are in Africa. Big road projects that cost billions in development aid in the 1960s and '70s were never maintained and are now crumbling.]

Back in 1985 at a Washington, DC, hearing on sustainable agriculture, the administrator of the U.S. Agency for International Development (USAID, or AID) acknowledged that it cost more to get fertilizer from an African port to an African farm than it did to get the material to Africa in the first place.

But that doesn't mean things are hopeless. We simply need to realize that new solutions are necessary. Solutions different from the ones imposed on these poor people so far.

First, planting huge fields of single crops of corn, of rice, of cassava—of anything—invites disease and pestilence. It gives native weeds like Striga an invitation to spread as never before, virtually building a high-speed roadway for weed expansion.

Second, chemical dependency is expensive, ecologically poisonous, and extremely damaging to the most important factor in any farming situation—the soil itself.

Third, disrupting native cultures with foreign ideas manages to be culturally condescending, sexist (men telling women what to do) and racist all at the same time. That's quite an achievement for something that's being done under the guise of charity.

And fourth, the neglect—or outright eradication—of native species that is a natural by-product of the Green Revolution and plantation farming goes beyond the bounds of simple cruelty to people of another culture, another color. It's a kind of genocide that eradicates every trace of the people: their plants, their culture, and in the case of famine, their very lives.

Such an approach is wrong. And it doesn't even work.

Fight Famine and Ease Population Pressures: Break the Chains of the Women of the World

Why do women farm in Africa, and why is it getting harder to farm? In *Africa in Crisis,* Timberlake explains:

> Men are forced to leave farms to find paid work in cities, work which often cannot be found.
>
> So women must labor on the farm, or go out to work on bigger farms to earn the cash the men are not sending home.
>
> Kenyan researcher Professor Philip Mbithi estimated that even when the man is on the farm, nearly half of his activities are not directly related to farm work.

And, as Timberlake points out, the environmental degradation that Africa has suffered has also greatly increased the workload of African women:

> As trees disappear, fuelwood sources move further from home. Women in northern Ghana may need a whole day to collect three days' supply of fuelwood, often walking eight kilometers with babies on their backs to… collect the wood.
>
> Soil erosion may also mean that women have to work more land or land further from home to grow food for the family, and declining water resources means they must walk further to collect water.

Like the degradation of the land, political and social changes that occurred as a result of colonialism also made women's lot worse in Africa. The two biggest enemies of women in this regard, explains Timberlake, were "the introduction of European legal systems and the advent of cash crops." The new European or colonial way of doing things caused much land to be placed under the ownership of men, even if that land had previously been a part of a woman's traditional family holdings. And the money to be made from the newly introduced cash crops especially "encouraged men to take over women's land rights," he adds.

Those cash crops, he notes, are an especially sore point with African women:

> Research around Africa has found that wives nearly universally oppose cash crops because they reduce the amount of land available for food. This would not matter so much if women were given a proportion of cash returns, but such money is usually considered part of male income.

In addition, remnants of the colonial way of doing things ensure that African women "get little help [with their farming] because extension workers and researchers direct all their attention to the men who are officially in charge of the farms." The men may not actually *do* anything on the farms, but they get the advice. Often, there's no choice. In many parts of Africa, there are cultural restraints against women being educated. And if the extension worker is a man, allowing him to speak with the women who actually grow the food is often unthinkable.

In a presentation to the House subcommittee on hunger in Africa back in 1985, Christine Obbo, a sociologist-anthropologist from Wheaton College who has studied African women extensively, explained that this sexist attitude is not part of traditional African culture. Like corn, plantations, explosive population growth, and famine, she agrees that this lack of basic rights for women is yet another legacy of colonialism:

> All available evidence suggests that what "yoked" African women [a reference to a previous speaker who said that African women's problem was that they were "yoked to a traditional culture"] was not traditional culture. ...
>
> [People] must stop blaming "tradition" and "culture" for which they have little understanding. Instead I suggest they focus on the real causes [of] the oppression of African women.
>
> [This] so-called traditional culture ... was a creation of Colonial domination which changed people's relationships to each other and how they did things.

Economic demands in the form of colonial taxes added to women's work burden, she notes. And European-style laws took away women's rights to land, money, and education.

A recipe for poverty. And Obbo notes that the goal of smaller families "will be rejected or ignored unless the major reason for large families—poverty—is shown to be ... controllable."

My good friend Gus Speth, president of the *World Resources Institute,* explained the situation this way at the committee hearings:

> Present laws and practices related to land ownership, credit and extension services mostly exclude women from access to loans and education, although women are an essential part of farming and food production in sub-Saharan Africa.
>
> In many African cultures women have traditionally played a major role in trade

as well as shared ownership of the land. However, the introduction of European legal systems has encouraged single ownership by men.

Moreover, recent migration by men seeking employment in urban centers has placed even more responsibility on women to sustain their families' food needs, and environmental degradation has further increased their workload.

Denuded landscapes and scarce water supplies mean vast distances must be covered and long hours spent gathering firewood and water.

Going over these thousand pages of hearings, I am impressed by the attitudes and ideas of many who came to testify. While reading one report on "agroforestry" (growing trees and other crops together) I was pleased to see that John Michael Kramer, a renewable natural resources program coordinator for CARE (Cooperative for American Relief Everywhere), simply referred to farmers as "she" or "her" throughout his testimony. I'm glad to see that some people are paying that much attention.

In the section of his report entitled "The Smallholder Farmer: Her Problems," Kramer explains that

It is difficult for western people to understand the depth … of the difficulties she must face every day.

Our farmer rises before dawn to start a fire for breakfast—a rough gruel and a glass of tea. She may then send her children or go herself to fetch water from the nearest well (often quite distant).

If she is running low on firewood, she will take a machete or small hand ax and, in the company of a group of neighboring women, set out for a day-long trip to gather dead branches and twigs.

If the fuelwood supply is ample, she may spend the bulk of her day in the garden or the fields tending crops with a homemade hoe. Near dusk she will return to relight the fire and cook a dinner similar to breakfast.

Too often her day is interrupted by the need to care for a sick child and, when severe illness strikes, a long journey to the nearest health clinic.

Because the land she tends is less than she actually needs to grow enough food for her family, she finds that, with time, her crop yields are diminishing. In order to produce enough grain to see her family through to the next harvest, she has reduced the fallow cycle … reduced yields follow.

Now she has to use the animal dung that once fertilized the field for fuel because she can't afford to spend the extra time needed to gather enough firewood.

Everyone [in the village] has the same complaints. She could move, but that would mean leaving the home of her grandparents and her relatives; her only social security.

She stays and hopes. If another drought comes she knows she may not survive, but what alternative is there?

Somewhere between 60 and 80 percent of all agricultural [product] that is raised throughout the world is produced by women … it is tragic to recognize that less than 1 percent of all agricultural land happens to be owned by women.

My good friend Albert Meisel, executive director of the League for International Food Education (which goes by the wonderful acronym LIFE), suggested in later testimony that the home garden system we endorse would help ease this economic imbalance: "Home gardens … provide a source of cash income for families. In at least one documented case, they have provided an alternative to villagers [leaving the farm] in search of wage labor." And these gardens can also help prevent the tragedy of families losing everything in a famine. As Meisel notes, the food—and cash—provided by "home gardens can reduce a family's debt to, and dependence on, traders during hard times."

Those traders are the ones who buy (at bargain rates) everything a starving family owns. So that when the famine crisis is over, even if the family has survived, it now has nothing. And that kind of crushing poverty, of course, is one of Famine's favorite friends. A return to home gardens really *can* help break this horrible cycle.

Meisel quickly adds that home gardening can do all this *without* adding to the burden of women. Just the opposite, he explains. LIFE's research found

> That gardening (unlike farming) is not in fact all done by women. … One of the beauties of [home gardening] is that it can be done … by people who are otherwise not productive and cannot add to the family budget; that is, older people, children, and sometimes even the men, curiously, because they take an interest.

Meisel explains that this help with the growing of extra food "adds to the resources" of African women and makes the family stronger. "And when the family becomes more self-sufficient, the woman becomes more powerful. That is my belief, in any case; and I think that belief is shared by a great many other people." I believe that as well. The challenge is in reaching the African woman and helping her achieve these advances.

As Peter H. Freeman, the geographer, reminded the House subcommittee:

> Women … have little power over rural development, although women are the major food producers, fuelwood collectors and users.
>
> Development planning is an urban, male, bureaucratic and technocratic activity [performed] by those who do not depend on the environment for their security.

The result, he explains, is an unending series of short-sighted, short-term projects that leave the land—and its women—in much worse shape than before.

And his reasoning is right on target. A system where men who are far away and who have little idea about the day-to-day work of farming and gardening, are nonetheless making decisions for the women who really have to do the work, is bound to fail.

The solution is obvious. Women must regain their lost status in Africa.

But there is also another, more subtle—and extremely effective—method that will help Third World people control their massive birth rate. And that is education and an increase in social status and basic rights for women. Throughout history and across the boundaries of civilizations, nothing has lessened the birthrate with more certainty than an end to the oppression of women. As subcommittee chairman James Scheuer put it during the hearings,

> Literacy [and] education is the best contraceptive, and it is essential for women … to accept the fact that there is a different role for [them]. …

That means that women must have access to education. They must have access to credit. ... There has been far too little progress in improving and enhancing the status and role of women, and until we do that, all of your garden production programs in the world won't be able to keep up with the ... population growth rate. ...

Steven W. Sinding, Ph.D., director of USAID's Office of Population, told the House subcommittee:

From my own experience in a number of Asian countries, and more recently in Africa, I have become personally convinced that until alternative roles for women are found they won't be interested in limiting family size.

From Marshall Green, a retired ambassador, speaking for the Population Crisis Committee:

Traditional African societies offer many cultural reinforcements for large families, and in many countries the virility of African men is still viewed as related to the number of children fathered.

A change in attitudes will only come ... when male perceptions of women cease to be focused exclusively on childbearing, child rearing and doing menial tasks. Strategies to improve the status of women are critical to this process, and AID [the Agency for International Development] should expand its Women in Development program.

AID should increase its support for innovative, small-scale projects... for women in developing countries in cooperation with [local] women's groups.

Small-scale projects. That's the key to helping women. And the key to lowering that staggering birthrate.

The simple truth is that every home garden, every alley-cropping success will allow women more time. More time to learn, more time to grow, more time to demonstrate their abilities and to understand that their value—their worth—in life is not limited to childbearing. And, slowly but naturally, the birthrate will fall.

Make no mistake, women in the Third World must be provided information about and access to safe and reliable contraceptives. Those slashed funds must be replaced. But woman-centered, garden-centered, locally controlled small projects—true development—will make smaller families a natural part of the chosen lifestyle of people who no longer live under famine's shadow.

Gaviotas

A Village to Reinvent the World

By Alan Weisman

Introduction

Way back in the prehistoric days of the early 1990s, when many readers of this book were preschoolers, I was a graduate student at the University of Virginia, working toward a PhD in Environmental Sciences. Our department emphasized the *science* in this, and quite consciously distanced itself from any form of "environmentalism," which had a bit of a stigma about it. I was allowed to take a course in environmental ethics, but was not given credit toward my PhD. There was no such thing as a major in "sustainability," nor was the word often used. I mention all this only to underscore the immense interest, relief, and perhaps even mystical yearning I felt when I heard a story about this bunch of crazy Colombian scientists and engineers who had taken it upon themselves to form a community in the middle of nowhere and try to solve some of the world's pressing problems. It was another decade or so before I actually encountered Alan Weisman's book, *Gaviotas*, but I always kept the story in the back of my head, hoping to find its origin. I am honored to be able to share this most moving of stories with you. In *Gaviotas* we see the most important, core principle for the restoration of this planet and the transition to human sustainability thereupon: *Integration*.

We have seen how there is a cascade of *dis*-integration, of degeneration, that has driven us to where we are today. Fossil fuels drive monoculture farming, which drives overpopulation, destroys soil, and dumps fertilizer on the land. The soil and fertilizer run into the oceans. The soil chokes reefs, destroying fisheries. The human waste from all the people combines with the fertilizer to create algae blooms. The algae die, rot, and suck all the oxygen out of the water, also destroying fisheries. Hungry people turn to crime and war, and further denude the landscape. The planet heats up, exacerbating all these processes. States fail, and chaos ensues. Yeats's oft-quoted poem about World War I perfectly describes our global eco-humanitarian crisis:

Things fall apart; the centre cannot hold;
Mere anarchy is loosed upon the world,
The blood-dimmed tide is loosed, and everywhere
The ceremony of innocence is drowned …

Gaviotas shows us that the *re*-integration of human and planetary systems is equally possible. Humans can rediscover the fundamental conservatism and decency of village life in a technical context. Scientist and rancher can work together. Energy production—even from biofuels!—can be reconciled with feeding humans and with restoring biodiversity. This is accomplished by working with nature rather than treating it as something to be occupied and conquered.

These are vague words. What are the specifics of this "integration"? The Gaviotas village is just one example of what many of us call *permaculture*, which stands for "permanent culture" as well as "permanent agriculture." Permaculture is a way of living that endures; it does so by taking account of local conditions and by integrating functions. A disposable water bottle is a bad thing, a source of pollution. At Gaviotas, however, a water bottle is built like a Lego block. Bottles can be filled with dirt and used to build a wall. A waste disposal problem (the bottle) has been integrated with a construction need (a brick). A monoculture tree farm is a bad thing, but at Gaviotas, a farm of sterile Caribbean pines is a bridge to restoring the hot, infertile soil of the Colombian *llanos* to a diverse and productive agroforestry complex that increases biodiversity and improves ecosystem services like cooling, while also providing food and fuel. When peat forests are burned off on Borneo for oil palm plantations, volcanoes of CO_2 form, making Indonesia one of the world's greatest global warming offenders. When these same oil palms are planted at Gaviotas, much more sparsely, with numerous other plants, we have a working, food-and-fuel producing ecosystem that can simultaneously support humans, the soil, and wild biodiversity. Yes, Gaviotas is a *designed* ecosystem; but as we have seen with Tikopia and other examples, humans have been designing functioning ecosystems for a very long time. Without such design, civilizations collapse. No global task is more important than learning to do this, on a global scale, but with attention to local details.

Overture

Years before Belisario Betancur became president of Colombia and proceeded to startle his fractured nation by risking a fledgling peace with Marxist insurgents who, at that time, ruled more territory than the government; before he filled the halls of state with works and recitals by Colombia's greatest painters, musicians, and poets, and invited the public in to see and hear; before he had the wizards from Gaviotas outfit his presidential mansion with their artful devices that coaxed the sun's bountiful energy through Bogota's dour skies—long before all that, he heard a story that he never forgot.

It was the kind of thing, he explained thirty-five years later to Paolo Lugari, the founder of Gaviotas, that jerked everything else into perspective. "It still does. Listen."

"I will, *Presidente*. And then I have one for you."

This was March, 1996: They were in Betancur's northeast Bogotá apartment, sipping chamomile tea. Outside, a cold rain pummeled the 9,000-foot skirts of the Andes. The roundfaced, silver-haired former president, now 73, sat in his leather chair, wrapped in a thick blue sweater and red wool scarf. Lugari, bearded and burly, evidently oblivious to the chill, wore his usual lightweight tropical suit. In his large hands, the china cup and saucer looked frail as eggshells.

"The year," Betancur began, "was 1962. I was a senator then."

A senator: Back then, the very notion had seemed miraculous. Belisario Betancur was one of twenty-three children born to nearly illiterate peasants. When he was eight, he'd found an illustrated volume of ancient history on his village school's bookshelves. Intrigued by the quaint pictures, he learned to read it. Soon he was scouring encyclopedias for more about the Peloponnesian wars, about Carthage, about the Roman emperor Hadrian, about anything Greek or Latin.

At his teachers' urging, his stunned parents eventually sent him to a seminary in Medellin, where he spent the next five years conversing solely in those classical languages—even on weekends, when Spanish was permitted, because he was routinely being punished for some breach of cloister decorum. His masters ultimately concluded that, however brilliant, he was too impetuous for the priesthood; the rector who expelled him arranged for his placement in a university. There he studied law and architecture, but ended up a journalist.

It was not an auspicious time. In 1948, Colombia had fallen into a horrific civil war; over the following decade, an epoch known today simply as *La Violencia,* hundreds of thousands died. There was little of comfort to report, but during those years Betancur discovered something of which most of his compatriots seemed barely aware: To the east of the

Andes, which bisect Colombia like a great diagonal sash, lay half the country, virtually uninhabited save for scattered bands of nomadic Indians.

The destiny that led him over the mountains took the form of a pilot who invited him to see exotic places seldom mentioned in the press. He went, and then returned as often as he could. What he found was Colombia's Amazon forest and, further north, *los llanos:* a vast savanna, drained by the Rio Orinoco, that stretched clear to Venezuela. Both were so huge and untouched that Betancur was soon convinced that, one way or another, the key to his country's future was there. Years later, in 1982, as a candidate for the presidency he would fly over the *llanos,* spot the community known as Gaviotas, land, and conclude that he'd been correct.

It took the first and only military dictatorship in Colombia's history, which began in 1953 and lasted four years, to finally snuff *La Violencia.* In its aftermath, Belisario Betancur, one of a scarred generation of survivors who had dreamed for an anguished decade of setting their country straight, entered politics.

"So there I was, a senator in a country trying to resurrect itself, having dinner in Washington, D.C., one evening at the Inter-American Development Bank."

At that time, 1962, the Inter-American Development Bank was a fresh offshoot of the World Bank, which had burst like a huge weed from the rubble of World War II and begun to broadcast its seed everywhere. The directors of the new multinational monetary funds were charged with cobbling together a battle-fatigued planet, by moving money into distant places where frequently the locals never before knew they needed it. Sooner or later, Betancur realized, these could include regions such as Colombia's Amazon forest and the *llanos.* His country needed development, he believed, but who would decide what kind? On his last visit to the *llanos,* a Guahibo Indian shaman had peered into a cloud of ritual tobacco smoke and correctly divined the precise arrival time of Betancur's overdue bush pilot. What did bankers at international lending institutions understand about such people and places?

That night over dinner, Bank president Felipe Herrera, a Chilean economist, told of a tiny Indian village on the high altiplano near Bolivia's Lake Titicaca, where he'd gone on a feasibility study for a proposed hydroelectric dam. Upon completing the site visit, his team realized they hadn't used their entire travel budget. Since the village lacked everything, they assembled the local chiefs and explained that they had some money left. In gratitude for hospitality and assistance, they'd like to give it to the community as a gift. "What project would you like us to fund here in the name of the Bank?"

The Indian elders excused themselves and went off to discuss this offer. In just five minutes they returned. "We know what we want to do with the money."

"Excellent. Whatever you want."

We need new musical instruments for our band.

"Maybe," replied the Bank team spokesman, "you didn't understand. What you need are improvements like electricity. Running water. Sewers. Telephone and telegraph."

But the Indians had understood perfectly. "In our village," the eldest explained, "everyone plays a musical instrument. On Sundays after mass, we all gather for *la retreat,* a concert in the church patio. First we make music together. After that, we can talk about problems in our community and how to resolve them. But our instruments are old and falling apart. Without music, so will we."

"And now," said Betancur, offering Lugari a silver dish of fried plantain slices, "lets hear yours."

"*Senor Presidente,*" said Paolo Lugari, shaking his head, "you're not going to believe this."

Juanita Eslava hadn't known whether to believe it either. The forest was enchanted, she'd essentially been told, by no less than Dr. Gustavo Yepes, director of the faculty of music at Bogota's prestigious Universidad de Los Andes. Juanita, grand-niece of a famous Colombian poet-composer, Luis Carlos Gonzalez, and granddaughter of a popular singer, was training at Los Andes to become a lyric soprano. She was on her way to a rehearsal for a 1996 European choral tour when she saw a notice on the bulletin board stating that a place called Gaviotas was looking for a few daring musicians.

"I don't know," she said, when Dr. Yepes explained that it was to help start an orchestra in a tropical paradise. "I'd have to miss Europe."

"Europe will be there next year. Its not going anywhere. When are you going to get another chance to do anything like this?"

That was hard to say, because Juanita had never heard of anything like this. For that matter, who had? Seriously, *los llanos?* Europe seemed closer.

She had at least heard of Gaviotas. That was unavoidable for any Los Andes student, because the road that climbed through the skirts of the cordillera toward campus wound right past the office that the Gaviotans maintained in Bogota. It was impossible to miss: an assemblage of brick and glass cubes, surrounded by colorful bursts of oddly graceful machinery rising above the eucalyptus. These included several windmills mounted on glossy yellow masts of varying heights, whose blades were not the typical narrow triangles but aluminum skewers, tipped with paddles shaped like cross-sections of an airplane wing. Alongside these stood a collection of bright red canisters of different sizes, assorted pipes and levers painted royal blue, and a bank of silvery rectangular surfaces. To passersby, the impression was technological yet pleasing and sculptural, like the promise of an appealing future waiting just beyond the encroaching urban bedlam below.

Engineering students at Los Andes knew about the silver rectangles, which had begun to appear around Bogotá during the mid-1980s, while Belisario Betancur was president of Colombia. Conventional wisdom said that solar panels wouldn't work in a city that was overcast more than half the year, but Gaviotas had come up with coatings for their models that gathered the energy even of diffused sunlight. Besides the presidential palace where Betancur formerly resided, their solar collectors were now atop condominiums, apartments, convents, orphanages, and on the brick edifices of Bogotá's 30,000-inhabitant Ciudad Tunal, the largest public housing complex in the world to use only solar energy to heat its water. The nations biggest hospital had not only converted their water heating system but had also installed solar "kettles" designed by Gaviotas technicians, capable of wresting temperatures sufficiently scalding from Bogotá's scant sunshine to purify water for drinking and sterilizing instruments.

But Dr. Yepes didn't even mention solar collectors to Juanita. He was talking about music. And trees.

Gaviotas wasn't just some high-tech research firm designing newfangled gadgetry, he assured her. Gaviotas was actually a place—a wondrous place in the middle of the practically treeless tropical plains of eastern Colombia, except it was now in the middle of a forest. An incredible forest of its own making. And now Gaviotas would soon be making music as well.

"*Música llanera?*" Juanita asked. If so, what did this have do do with her? The traditional country music of the Colombian *llanos*, with its harps, four-string *cuatros*, and twangy *bandolas*, was a long way from the Italian arias she sang.

Gustavo Yepes explained. One evening a few years earlier, he had been introduced to Paolo Lugari after a choral performance of Bach's sacred music. That night, Lugari had pumped Yepes's hand and boomed in his basso profundo, "Tell me, Gustavo: How do composers' creative passions, which are born of random, nonlineal emotion, deal with the structure of music, which is mathematical and therefore lineal?"

It was a strange, remarkable question, but Yepes had heard that this was a strange, remarkable person. "I imagine it's much the same," he replied, "as what happens at Gaviotas. People who dare to build a Utopia use the same materials available to anyone, but they find surprising ways to combine them. That's exactly what composers do with the twelve tones of the scale. Like you, they're dreamers. In a dream you aren't limited by what is assumed to be permissible or possible."

"Gaviotas isn't a Utopia," Lugari interrupted. "Utopia literally means 'no place'. In Greek, the prefix 'u' signifies no. We call Gaviotas a *topia,* because it's real. We've moved from fantasy to reality. From *utopia* to *topia.* Someday you need to come see it."

That day, Yepes told Juanita, had unexpectedly arrived in October, 1995. Paolo Lugari had called to say that some German journalists had chartered a plane out to the *llanos* to see Gaviotas. There was an extra seat, and he especially wanted Yepes to accompany them.

"Why me?"

"You'll see."

What he saw—and heard—belied Lugari's protestations; to Yepes, Gaviotas seemed not only proof that Utopia on earth was possible, but that it was arguably more practical than what currently passed for conventional society. Five hundred kilometers away from his increasingly frightening city, Yepes had found himself in a tranquil village, shaded by the gallery forest of a tributary of the Rio Orinoco and filled with flowers and dazzling, melodious birds. The people of Gaviotas collectively exuded a quality so novel that Yepes wasn't sure he'd seen it before—but once encountered, it was unmistakable: They were happy. They rose before dawn, worked hard and productively, ate simply but well, and were peaceful. The machinery they used dominated neither them nor their landscape: it was mostly of their own design or adaptation, and mostly quiet. "May I retire here?" Yepes had asked Lugari, after watching children playing on a seesaw that was also a water pump, which tapped kid power to replenish a reservoir for the Gaviotas school.

"Don't wait to retire. Come sooner. You're exactly what we need."

They were walking down a red dirt path that led past a grove of mango trees, an outdoor basketball court, polygonal modular living quarters, and a community meeting hall with a parabolic swoosh of roof, contoured from shining metal to deflect the equatorial heat. Just south of town, the path widened into a road, with a tall pine forest rising on either side. They exchanged waves with six men and a woman dressed in caps, colored neckerchiefs, tee shirts and tool belts, who rode past on thick-tired bicycles. Lugari steered Yepes into the forest as he began to explain. "For the past quarter-century—ever since Gaviotas began," he said, "I've been studying the history and literature of utopic communities."

"I thought you said this wasn't Utopia."

"Neither were any of those other places. They were attempts." Lately, Lugari had been reading about the famed experiment of 17th century Paraguay, when Jesuit priests arrived to evangelize the New World. Until then, colonizers throughout most of the Americas had considered indigenous peoples either expendable savages or exploitable slaves. But the Jesuits who ended up far from the trade routes, in the distant region where the borders of Brazil, Argentina, and Paraguay now converge, saw the resident Guarani Indians as a kind of *tabula rasa*: untainted *Homo sapiens* in their natural state, potentially perfectable. Being missionaries, of course, meant having certain preconceptions about perfection, and these Jesuits soon set about replacing the natives' language, god, and means of sustenance. Their missions, aptly named "reductions," were consummately paternalistic but nevertheless benevolent, self-sustaining communities that prospered for more than a century, until the Jesuits fell into disfavor with Spain and Portugal and were expelled from colonial Latin America.

Paolo Lugari was not interested in evangelism— Gaviotas didn't even have a church. What enthralled him about that historic Paraguayan experiment was the music. "Everyone," he told Yepes, "was taught to sing or to play a musical instrument. Music was the loom that wove the community together. Music was in school, at meals, even at work: Musicians accompanied laborers right into the corn and *yerba mate* fields. They'd take turns, some playing, some harvesting. It was a society that lived in constant harmony—literally. It's what we intend to do, right here in this forest. That's why I asked you to come,"

But Yepes wasn't listening—or rather, he *was* listening, but not to Lugari's words. He stopped and held up his palm. "Quiet for a moment," he said. Silence, except for the drumming of a woodpecker and a rustle of breeze in the pine boughs. Then: "Keep talking," he whispered.

"What?"

"Did you hear it?"

"Hear what?"

"Talk."

They were in a thicket enclosed by forty-foot Caribbean pines and a leafy tangle of decidous trees and shrubs. Even at tropical noon, the forest air was delectably cool. Amidst the profuse foliage of the understory, it was hard to discern that the pines were actually growing in evenly spaced rows. Thirteen years earlier, this woodland—now the biggest reforestation in Colombia, more extensive than all the government's forestry projects combined— had been mainly empty savanna, devoid of anything but low, nutrient-poor grasses. By 1995, the number of trees Gaviotas had planted was approaching six million.

Yepes was taut with excitement. "Paolo. Just say something. Anything. "

Shrugging, Lugari started to explain how, when he and the early Gaviotans first came out here from Bogota in the early 1970s, they had tested hundreds of crops, but nothing thrived in these highly acidic, leached tropical soils, whose natural levels of aluminum bordered on toxicity. Then, a Venezuelan agronomist seated next to him at a conference in Caracas suggested trying tropical pine seedlings obtainable from Honduras.

The trees grew. The Gaviotans debated among themselves whether it was wise to cultivate an exotic species. Some argued that the issue was political, not environmental, since the same pines also grow in Panama, which was once part of Colombia. Had the United States not stolen the isthmus and installed a puppet government in order to dig their canal, these would still be native Colombian trees.

The controversy, along with the matter of what to do with pines since they weren't edible, was settled by a succession of random occurrences, the kind of unpredictability that the Gaviotans had come to love as they tinkered with improving reality. Who could have guessed that Caribbean pines would prove to be sterile in the *llanos,* posing no invasive competition to local flora? Who could have known that their bark resin, a natural protection against the tropics' array of hungry insects, would flow so copiously here that it could be harvested like maple syrup—more, really, like milk from cows, because tapping the thick amber liquid seemed to stimulate production without hurting the trees? Or that here pines would mature nearly a decade faster than forestry texts predicted? Of that until a few months ago, Colombia had been importing millions of dollars' worth of resins annually for paint, varnishes, turpentine, cosmetics, perfume, medicines, rosin for violin bows—until, that is, Gaviotas inaugurated a forest products industry that involved leaving trees in place, not mowing them down?

And, most wonderful of all, Gustavo, who could've—"

"Wait."

I was just getting to the most important part.

"Did you say violin bows?"

"Right. That's one of the reasons I wanted you to come here. But not just rosin. We realize that when we have to thin the forest we can use the surplus wood to start a musical instrument factory, and—"

"Do you have any idea how perfect a place this is to make music?"

"Exactly. That's why we wanted you to come here."

"No," Yepes insisted. "You don't know what I mean. Listen."

So Lugari did, and that's how, three months later, Juanita Eslava found herself not in Paris, but in the middle of a forest under a full moon at midnight, in what most of her compatriots considered the middle of nowhere, preparing to sing an aria by Respighi. According to what Yepes had told her, yet another random stroke of luck had inexplicably imbued the Gaviotas forest with magnificent acoustics. "We were standing in the woods," he recalled, "and suddenly I realized that I could hear distant voices, as though they were amplified. I clapped my hands. Then I yelled. I made Lugari whisper. There's incredible resonance in there. We don't know why. Maybe the forest canopy vibrates. Maybe it has to do with the physics of unorganized spaces. Paolo wants an engineering student to write a thesis about the effect. I just want to build a bandshell to focus it."

Like a pair of excited kids, right there Yepes and Lugari had started planning an outdoor amphitheater-in-the-pines, with some form of retractable roof for rain, like the one on the Gaviotas administration building. "Probably need to encase the whole thing in mosquito netting, too," Paolo had added. They envisioned concerts of classical symphonic instruments and also a resident *llanos* orchestra, comprised of entire sections of Orinocan *cuatros, bandolas,* and harps made from renewable Gaviotas pine.

Juanita wasn't so sure about these ambitious schemes: Rather than forty *bandolas* plucking Beethoven's Sixth, she preferred the idea of combining violins and cellos with folk instruments to create a sonorous new mix of timbres. She was impressed, though, at how serious the Gaviotans were about their musical future. During the 1970s and 1980s, when many of its famous technological innovations were being developed, Gaviotas entered agreements with Juanita's university and several others to bring scientists and engineers here to research their graduate theses. Under the most recent accord with the Universidad de Los Andes, however, Gaviotas had requested painters, sculptors, and musicians. "There's no such thing as sustainable technology or economic development without sustainable *human* development to match," Lugari had told her when she arrived. "Over twenty-five years, Gaviotas has accomplished much, but we need so much more."

Juanita's mission was to establish a classical music program in the Gaviotas school: the first step toward building an orchestra. She was also to get to know, and record, resident Gaviotas *llanero* musicians. And, finally, she was to tramp around the woods until she found the spot where her voice projected best, so the Gaviotans would know exactly where to build their theater—if, in fact, this business about its alleged acoustic properties were true, and not simply Yepes' imagination having been seduced by the spell of the place.

So there stood Juanita Eslava, her long dark braid glinting under lemony moonlight in a forest that her distinguished professor swore had magical properties, ready to find out. For some reason, she had delayed this moment until now. Maybe it was because Gaviotas had turned out to be such

an island of blessed tranquility in the midst of her roiling nation. During her first month here she had learned as much as she taught, from listening to *musicos* who could echo the gallop of horses on their *bandolas* and the sweetness of the trade winds on their harps. Every morning, she awakened to a delirious symphony of nesting tanagers, cotingas, and oropendolas outside her window. The Gaviotas schoolchildren she taught to sing were the healthiest humans she'd ever met, happy and unafraid as the monkeys cavorting overhead. Everything was so sublime that maybe she was scared to spoil it by putting something she suspected was implausible to the test. But on this night of a full moon, a group of her new friends finally had dragged her off to sing in the trees. They spaced themselves at varying intervals: ten, twenty, fifty meters away from where she stood. Then they waited.

Juanita struck a tuning fork against her knee, hummed the pitch, closed her eyes, and inhaled deeply. All around her was lush, fragrant evidence of an indisputable miracle, a portent that the place might very well be enchanted. In the moist, sheltered understory of the Gaviotas pines, an indigenous tropical forest was regenerating. A team of frankly amazed biologists from Colombia's Universidad Nacional already had recorded 240 species probably not seen in the *llanos* for millennia, except in fragments of terrain alongside the streambeds. Another unpredicted stroke of fortune had rendered moot the concerns about introducing a monoculture of *Pinus caribaea* into the *llanos*—it was as though the savanna's thin green ribbons of riparian forest had overflowed their banks and were spreading across the plain.

Some trees, such as the slender purple jacaranda Juanita was leaning against, already towered higher than the pines. With thousands more hectares available for planting, the Gaviotans had decided to let the native species slowly choke out the *Pinus caribaea* over decades and return the *llanos* to what many ecologists believe was their primeval state: an extension of the Amazon. Already, the populations of deer, anteaters, and capybaras were growing.

When she opened her eyes and began to sing, an angel's aria from Respighi's *Laud Per La Nativity Del Ignore* emerged.

> *Pastor, voice che vegghiate*
>> Shepherds, you who watch over
> *sovra la greggia en quista regione;*
>> your flocks as they graze here;
> *i vostr'occhi levate*
>> lift your eyes

By the calendar, this was just before the March equinox, but Juanita had spontaneously decided to invoke Respighi's celebration of the Nativity. Her voice, hesitant at first, began to billow through the forest like a silver mist, expanding as it swirled from tree to tree. Nightjars, owls, and lapwings joined in, cooing in plaintive dissonance that resolved into haunting harmony as she continued:

> *ch'io son l'Agnol de l'eternal magione.*
>> for I am the Angel of the eternal mansion.
> *Ambasciaria ve fone*
>> I bring you a message,
> *ed a voie vangelizzo gaudio fino*
>> and news of pure joy

Celestial music rose into the branches. The forest canopy gathered and magnified her clear tones, showering them down around her friends like gently falling pine needles. When she finally ended, they gathered around and embraced her, several nearly in tears. Luisa Fernanda Ospina, the bacteriologist in charge of quality control at the resin factory, stared in awe at the trees rising moonward. "This place is proof that God exists," she declared.

Gonzalo Bernal nodded. During the 1970s and early 1980s, he had directed the Gaviotas School;

now, in the 1990s, he was community coordinator. "Now I know for certain that we live in paradise," he whispered. "We can hear angels."

"So now Gaviotas will become a choir of angels. When I went there the first time," Belisario Berancur reminded Lugari, "I merely saw prophets. But I have a weakness for prophets like you who preach in a desert. It was like I'd heard a message. I immediately wanted to convert all Colombia into a Gaviotas."

He leaned back and gazed at a pair of framed sketches above the gray velvet couch opposite his chair, landscapes of the Colombian Andes. "Imagine," he sighed, "if this were all Gaviotas."

The sketches, signed and dedicated to him, were studies for oil paintings by the Colombian master Alejandro Obregón, one of which now hung at the United Nations, the other in the Vatican. Over the bookcases were more works by Colombian artists, gifts for the presidential palace that were later removed by Betancur's successors. The most famous of these, a painting that became the symbol of his presidency, occupied the space over the mantel. It showed a plump white dove with a fig leaf in its beak, portrayed by the renowned Colombian painter and sculptor Fernando Botero.

In the 1980s, its likeness had been borne aloft by exhilarated throngs marching through the streets of Bogotá, Cali, Medellín, and Cartagena. Botero's dove adorned posters for concerts, banners for theater festivals, children's clothing; it became the embodiment of the hope engendered by Betancur's peace initiative. While in office, he had proposed an unprecedented amnesty to thousands of Marxist rebels who had formed guerrilla armies a few years after the 1957 truce that ended *La Violencia* supposedly brought peace to the land. This new uprising, which had killed many thousands, was still underway and was now the longest-running armed insurgency in Latin America. Under Betancur's plan, guerrillas could trade their weapons for the chance to create their own political party and battle legitimately within the civil system. The largest insurgent army agreed to participate, and, in 1984, scores of guerrilla soldiers laid down their arms. Subsequently, the party they and their sympathizers founded, the Patriotic Union, won elections nationwide for mayoral seats, town councils, and for Colombia's national congress.

Within a decade, the majority of those victors—some two thousand, plus two presidential candidates—had been assassinated. The perpetrators, who sometimes issued gleeful press releases, were right-wing paramilitary death squads.

The guerrillas, of course, retaliated. Soon, their attacks and ambushes surpassed former levels, as did kidnappings for enormous ransoms to finance their subversion. In a monstrous reprise of *La Violencia,* massacres of civilians whose villages were supposedly aligned with one side or the other became almost weekly events. These atrocities were blamed on both right-wing paramilitaries and left-wing guerrillas, but rarely solved. Both extremes had become so deeply corrupted by the bounty of the narcotics trade that, after a while, it barely seemed to matter which was which.

Faster even than the Gaviotas pines, cattle ranches of drug lords had advanced across the *llanos,* in cadence with the march of coca cultivation in the Amazonian provinces to the south. By 1996, the current presidential administration was so tainted by a scandal involving drug spoils that several prominent members of the president's party and campaign staff—including Fernando Botero's own son, the ex-Minister of Defense—were in jail. When a massive Botero sculpture of the peace dove was wrecked by a bomb that killed dozens in a Medellín park one Sunday, the grieving artist directed that it be left in pieces as a monument to the shambles his country had become.

Approaching the end of the century, Colombians frequently had come to wonder aloud whether their nation could actually survive. "These things take time," Belisario Betancur would remind people. "I never thought that the process we began would be completed during just one presidential term. Over three or four decades, peace was systematically destroyed in our country. It was like an *ovillo,* a ball of wool that had been unraveling for years. To pretend that one could roll it up in four years would have been an illusion. But we had to start somewhere."

During Betancur's time in office, the *llanos* became the haven to which he retreated so often that his own party leaders complained, because there were no votes out there. "Not many votes, but so much Colombia," he'd reply. The boundless landscape restored his spirit, and Gaviotas—where his plane often touched down unannounced—was where he could happily stand in line for meals like everyone else, surrounded by people who, since they lived contentedly without any government themselves, embraced him and not his office.

"The history you are writing reads like poetry," Betancur said, as he and Lugari embraced at the door. "And now you're setting it to music, too."

"You will join us for the first concert," Lugari replied. "It will be in your honor."

The old ex-president beamed at the thought of returning to Gaviotas.

"This," he once told Gabriel García Márquez when he sent him there, "is what Colombia needs."

"This," he'd repeated, as President Felipe González of Spain and his family boarded a plane bound for Gaviotas' grass airstrip, "is what Latin America needs."

And when a group from the Club of Rome visited Gaviotas in 1984, Club founder Aurelio Peccei declared to Betancur, "This is what the world needs."

Afterword

I first learned of gaviotas in 1988, during a *New York Times Magazine* assignment to write a 4,500-word "portrait of Colombia." At the time, I'd been in the country for two months on a Fulbright research grant, long enough to know how daunting this would be. It would have been easy enough to serve up a bloody slice of the current cycle of narco-bombings, paramilitary assassinations, and guerrilla retaliations. But the true sorrow of *la guerra sucia*— the ongoing "dirty war"—was that Colombia was so much more than its numbing headlines, something that everybody seemed to have forgotten.

It was especially much more than the coffee and coca plantations that people in my own nation imagined. Unlike many other struggling lands relegated to that lowly caste known as the Third World, Colombia was hugely blessed, both with resources and skilled people to utilize them. Its literacy rate rivalled or surpassed that of most countries on earth, including my own, and dozens of fine Colombian universities produced brilliant scientists, engineers, writers, technicians, and business leaders. Colombia boasted seasoned industries ranging from textiles to book-binding, more than a hundred exportable crops, truly vast mineral deposits (coal, oil, and emeralds), enough fresh water to rank third in world hydroelectric potential, and possibly the planet's richest ecosystem.

This last fact intrigued me so much that I asked the director of Colombia's national parks to tell me which place best exemplified his country's astonishing biodiversity. That, he replied, would be the Serranía de la Macarena, Colombia's oldest nature preserve, the 80-mile-long geologic uplift whose forests and rivers were believed to harbor more species than any similar expanse on earth. It was just a few hours southeast of Bogotá—"But," he warned me, "you can't go there. It's too dangerous."

He explained that the jungle canopy of this incomparable world treasure concealed a major guerrilla command base, around which a growing number of coca growers were steadily replacing large swathes of this ecological Eden with a monoculture of wispy, lime-green shrubs. Of course, I had to go. The Macarena seemed the perfect symbol to portray a nation extravagantly endowed but also savagely cursed—a nation that I increasingly recognized as a microcosm of all that simultaneously plagues and holds promise in our bountiful, beleaguered world.

Through contacts arranged by a helpful Colombian correspondent for a Soviet news agency, I was able to negotiate safe conduct from the Marxist-Leninist *Fuerzas Armadas Revolucionarias de Colombia,* who occupied the Macarena. It took a day's bus ride plus two days on foot to reach their impressive bamboo encampment. After two more days of politely chatting political doctrine with rebel *comandantes,* I was led by guerrilla guides on a journey through unparalleled biological splendor, which was only interrupted by the coca fields we crossed— and by a sudden military ambush that wiped out most of a forty-man FARC detail, moments before we were to rendezvous with them. My last view of the Serranía de la Macarena was over my shoulder, running from helicopter-borne machine-gunners and squadrons of Colombian Air Force T-33 bombers that were reducing yet another chunk of paradise to smouldering ashes.

Soon after filing my report, I left for home. "Someday," said the journalist whose connections had enabled my glimpse of the Macarenas beauty and anguish, "you must return to write a hopeful environmental story." Still shaken by what I'd just witnessed, I couldn't fathom what hope I might discover in this tortured land. Then she told me about a remarkable community that, for years, had ignored surrounding strife to prosper in a most unlikely setting: the bleak eastern Colombian *llanos*. It was called Gaviotas.

I recognized the name: One day during my research I had encountered technicians whose caps bore the yellow-and-green Gaviotas logo, turning rooftops in a Bogotá slum into thriving hydroponic gardens. Years would pass before I saw them again. From 1990 to 1992, with three other journalists I produced a 23-part series for National Public Radio that documented how so-called progress was turning entire traditional cultures into endangered species, often by literally ripping the ground from under them. Set in a dozen developing countries, our series, *Vanishing Homelands,* described the threat humanity now posed to its own habitat. It ended with me gazing at Antarctica's ozone hole and speculating whether the entire planet—everyone's traditional homeland—was now at stake.

These sobering reports led to *Searching for Solutions,* a public radio series that explored possible antidotes to what currently ails the earth. Over two more years, in places such as Brazil, India, Europe, the Middle East, and our own United States, we chronicled attempts to produce enough food and energy for straining populations—and how to humanely check the growth of those populations—without sacrificing nature and culture in the process. Mostly, we discovered how complicated the solutions will be. But the most heartening program of the series turned out to be the one I brought back from, of all places, Colombia.

To get that story, I traveled for sixteen hours in February, 1994, from Bogotá to Gaviotas in a Diahatsu jeep, including delays obliged by official army roadblocks and by truckloads of armed men whose affiliation was less certain. The car and body searches proved a relief from the sensational pounding of the highway that never was. Being the dry season—the road is often impassable the rest of the year—I was so coated with powdered clay that one sergeant doubted whether my passport picture was really me.

I was accompanied by my journalist friend from Bogotá who had formerly worked for the now-defunct Soviet news agency. I was traveling to Gaviotas to see sustainable technology created by and for the Third World; the compelling interest for my companion, who had seen loved ones and journalist colleagues massacred, was Gaviotas' reputation as an island of hope amidst Colombia's ongoing tragedy. We suspected there might be a connection.

During our stay, she was moved to tears that such a peaceful refuge could exist in her nation. But the story of Gaviotas apparently also touched a profound yearning in my own First World country: To this day, I hear from people who listened to my NPR documentary, or from readers of a subsequent article I wrote for the *Los Angeles Times Magazine,* who wish they could live in Gaviotas or start their own in the United States.

I have since returned twice, once overland, and again, during August, 1996, in Pepe Gómez's old single-engine Piper Dakota, now owned by a neighboring rancher. That flight was advisable because much of the road lay submerged during that torrential time of year, and because sections of it—especially the stretch across the Andes between Bogotá and Villa-vicencio—were under siege by guerrillas. Since then, Colombia's political pains have not eased; nevertheless, Gaviotas continues to advance. The resin factory's boiler, fueled by culls from their own forest, has been tuned successfully to emit no visible smoke. Its co-generating two-cylinder steam engine, now installed, portends to be so efficient that the diesel plant that long augmented the ten-kilowatt micro-hydro turbines can finally be junked, making Gaviotas at last self-sufficient in energy. As a result, Gaviotas was awarded the 1997 World Prize in Zero Emissions from ZERI, the United Nations' Zero Emissions Research Initiative.

That same year, in response to a nationwide scarcity due to ever-increasing demand for that singular Colombian delicacy, edible colony ants, Gaviotas designated a portion of its savanna as a preserve for this species of six-legged wildlife.

In another, more surprising move toward complete sustainability, Gaviotas has elected to sell its cattle herd and apply new, modular techniques for raising rabbits, chickens, and fish. These systems, more efficient than husbandry technologies available back when the Peace Corps was in residence, will be run as private enterprises by Gaviotans, in what they hope will be a healthy economic mixture with their cooperative forestry and resin industry. "It also exemplifies our recognition," Paolo Lugari told me, "that too much red meat is bad for us, that too many cow pastures are bad for the environment, and that too much *hamburger-ización* is bad for the world."

The world: It has come a long way since my friendship with a Colombian journalist who worked for the Soviets had her colleagues trying to convince her that I was a C.I.A. agent, and had the U.S. Embassy warning me about consorting with Communist spies. Yet the fading of the Cold War has revealed clearly that a far more incandescent and protracted battle—a potentially apocalyptic resource war—has been stealthily gathering intensity throughout the latter part of the twentieth century. Despite sustained efforts to mobilize all human wisdom and will in defense of nature and sanity; we have yet to quench the flames that consume our forests, or to dampen the greed that stokes our excesses.

Yet a place like Gaviotas bears witness to our ability to get it right, even under seemingly insurmountable circumstances. With these pages I have returned to the inspiration Gaviotas embodies to replenish my own hopes. May we all journey there, again and again, to bring its promise home and spread it afar.

Postscript to the 10ᵗʰ Anniversary Edition

The newly planted palms, only a dozen, were maybe four feet high: a single line of green pinnate fronds spaced thirty feet apart. So far, it didn't look like much, but Paolo Lugari was positively beaming as he hopped over to inspect, the tip his cane punching swirls of red from llano soil that had been broken only the day before. "How many will you plant per hectare, Oto?"

"Just forty-three," replied Otoniel Carreño. "A hundred fewer than in a monocultured plantation. Gives the seedlings plenty of sun and space to grow—and room for company later."

"*Excelente*," boomed Paolo. "The agriculture of the future will be the art of taking advantage of light. And," he added, brandishing his aluminum cane at the impressive botanical display rising around us, "it will be a polyculture."

We were standing in the forest that had bloomed in what twenty-five years earlier had been a monotonously treeless plain. More than a decade had passed since I'd last been here. Otoniel, his moustache grayer but still looking trim in jeans and a white shirt with the Gaviotas yellow-green logo on the breast pocket, was a ruddy testament to a healthy life led outdoors. Paolo, now in his mid-sixties, seemed indefatigable as ever, although slowed somewhat after shattering several bones in his foot on a slick Bogotá street a few years earlier. But if they were aging on a normal timescale, the Gaviotas forest had assumed a pace all its own.

It was nearly half again as big as when I'd last seen it: expanded by another three thousand hectares. Most impressive, however, was how native foliage was swallowing the original pines. A riot of jacarandas, ficus, *yopos,* monkey pods, *tunos blancos,* curate, laurels, and various ferns all but concealed the neat rows of *Pinus caribaea* among them. Although a plantation, it looked far more like a wild forest. And it was: not only deer, anteaters, and capybaras lived here, but tapir were frequently spotted and even an occasional puma.

Were the pines being overwhelmed, I wondered, by competition from primordial species that had sprouted in the shelter of their understory?

"On the contrary," Otoniel replied. "The mix of plants has just made the soil better. These pines were planted in 1983. Last time you were here, we were tapping them for resin. They've kept growing—some are more than thirty meters[1] tall. They're so robust, they're ready to be tapped again."

Except the Gaviotas resin crews were currently miles away, harvesting liquid amber resin from mature trees I remembered as seedlings smaller than these palms. It might take years before they cycled back to this parcel, Otoniel said. Meanwhile, they were trying something new here. Using a bladed roller they'd specially designed, they had mowed and mulched a swathe of native underbrush, spaded it into the soil, and planted African oil palms. They were betting that, like their astonishingly productive pines, this commercial crop—now cultivated throughout the world's tropics for cooking oil, and lately for biofuel—would grow far better among other plants that contributed natural nutrients to the soil.

It was definitely more natural than the artificially fertilized monocultures I'd recently flown over en route to Gaviotas. From the eastern skirts of the Andes, I'd seen thousands of hectares of African palm plantations filling what had been cattle pastures a decade earlier. But were non-native African palms—and, for that matter, biofuels—a good idea? Weren't tropical forests and sustenance farmland, from Indonesia to Africa to Colombia itself, being lost to exotic energy crops at a shocking rate?

"*Sí*," agreed Paolo. "There's no justification for displacing one square centimeter of native forest for biodiesel. Nor of food production. First come

mouths, then motors. But that's not what we're doing here."

The difference, he said, was that virtually no trees, let alone food crops, prospered in Colombia's rain-leached eastern savannas until Gaviotans learned to cultivate *Pinus caribaea* here. When a native forest sprouted in their shade from seeds brought by winds, birds, and animals from jungles along Orinocan streams, they'd discussed trying coffee and rubber trees in the replenishing soils among the pine rows. Then in 2003, while Lugari was in Boulder to give a speech, he met University of Colorado engineers who proposed a Gaviotas biodiesel project.

They'd taken him to see a biodiesel plant that used vegetable oil and recycled restaurant grease. The technology seemed straightforward. "I'm sure we can tropicalize this," Paolo declared. Colombian land barons were already raising African palm for the rich yields of edible oils from its fruit and kernels. Why couldn't it work for fuel?

A year later, a team of Colorado volunteers arrived. In three weeks, they and the Gaviotans built what, as far as anyone involved knew, was the world's first biodiesel plant that used palm oil.

They did not build it at Gaviotas, however, but at the Bogotá factory where Gaviotas made solar collectors, windmills, and pumps. This was at a time when civil unrest in Colombia had escalated ferociously. The llanos around Gaviotas were considered too perilous for foreign visitors in a country where kidnapping had become a major fund-raising activity for both left-wing guerrillas and right-wing paramilitaries. With ransoms for North Americans sometimes exceeding a million dollars, their presence would have made defenseless Gaviotas even more vulnerable.

The eastern savannas' notoriety as a no-man's-land where lawless bands roamed was the main reason Gaviotas hadn't grown as planned since I'd last been here. Rumors abounded of civilian

---

1 about 100 feet

massacres and war taxes extorted or vehicles seized by one group or another for their respective political causes. It was difficult to know what was exaggerated and what was true. Some of the worst confirmed violence was in the oil-rich provinces of Arauca and Casanare to the north of Vichada, but the entire llano was considered dangerous.

Perhaps because Gaviotas remained famously unarmed, no one there had been harmed by the mayhem that bedeviled other parts of the country. Nevertheless, they sold the trucks that once brought their pine resin to market, and hired contract transporters who charged accordingly for the risk of traveling the treacherous road between Vichada and Bogotá.

As the new century turned, one again the ongoing tragedy of Colombia displaced huge populations: more than two million internal refugees, at the time second only to Sudan. At least 1.5 million more fled the country altogether. With human rights workers, journalists, union leaders, and even innocent TV personalities being assassinated—the last presumably because their celebrity focused attention on the perpetrators' demands—Gaviotas kept as low a profile as possible. Postponed along with the proposed satellite village of Odisea was a Gaviotas Internet site: the less visibility the better, all agreed.

Yet survival still required making a living, often a challenge when clients for Gaviotas' pine resin were ensnared by fiscal crises that deepened along with the civil disorder. "It has taken," said Paolo dryly, "a great deal of imagination."

And flexibility, a trait at which Gaviotas fortunately excelled. Hence, the biodiesel plant: a collection of 5,000-liter vats, pipes, and galvanized tanks. I'd toured in Bogotá before we'd boarded a single-engine Cessna for a visit to Gaviotas. It produced high-quality vegetable diesel from palm oil they bought from local growers—an achievement which, like some other Gaviotas experiments over the years, was impressive but not necessarily profitable.

Lugari shook his head, chuckling. "When the group from Colorado came, we could get crude palm oil for $450 per ton. That was before the world discovered that hydrogenated oils are a bad idea." Suddenly, industrial food processors began snatching up palm oil—which, though highly saturated, contained no trans fats. "In the past three years, the price of edible palm oil nearly tripled. Economically, refining it into biodiesel makes no sense."

Nevertheless, as he and Otoniel were now telling me, given the disastrous costs of atmospheric carbon dioxide, renewable biodiesel from palms grown where nothing else normally grew anyway might make sense after all—especially in a world where petroleum cost more daily. As if to underscore that point, a grizzled Pompilio Arciniegas, the government forester who, I was pleased to see, had still never left Gaviotas, rode up on a motor scooter.

"I thought this was strictly bicycle country," I said, shaking hands.

"No longer possible with a tree farm this big," Pompilio replied.

They'd built several gossamer, light-weight fire lookout towers of steel lattice anchored with guy wires, which were manned continuously. But for a fire-fighting crew to respond to an alarm by bicycle in a forest this extensive would be suicidal.

"By far, Gaviotas' biggest success of the past ten years," said Paolo, "is keeping thousands of hectares of pure fuel from burning down."

Their mechanized needs went beyond fire prevention. Gaviotas had stayed alive by becoming an agro-industrial cooperative, and the industry part meant tractors, mulchers, plows, and disks as well as motor scooters. Their biodiesel factory in Bogotá could produce enough to run them all. But rather than keep buying costly crude palm oil to refine, they'd calculated that with thirty hectares of fast-growing African palms planted in the fertile soil between their pine rows, in a few years they could produce all their own. "We'll be completely self-sustaining in fuel—self-sustaining *and* non-polluting." Paolo said. "And we'll have enough oil left over for cooking."

It was an ambitious plan, and it had already spawned another even bigger. ZERI—the same international Zero Emissions Research and Initiatives foundation that had awarded Gaviotas its world clean energy award in 1997—had approached the Colombian government. In Vichada and the neighboring province of Meta alone, there were millions of empty hectares similar to the terrain around Gaviotas. Why not plant them in pines, palms, and whatever else nature added, to capture carbon dioxide and to produce clean, renewable diesel for the whole country?

The government was interested. Soon ZERI founder Gunter Pauli, who as a young man had accompanied Club of Rome founder Aurelio Peccei to Gaviotas, was being flown with his staff and key guests by Colombian Air Force officials to see what Gaviotas had done. From there, they traveled to inspect Marandúa: the same 70,000-hectare military preserve on the Río Tomo, halfway to the Venezuelan border, where former president Belisario Betancur had once dreamed of starting a Gaviotas forest writ large, to resettle thousands of displaced Colombians from the nation's overwhelmed cities and employ them in a new Colombian capital of the plains.

In a country as politically complex as Colombia, whether the dream will actually materialize this time depends on multiple factors, not the least of which is securing financing for what could be the biggest sustainability project in the world. But the very fact that the dream had not died when horrific national events overwhelmed it two decades earlier heartened me. And what Paolo Lugari took me to see next further confirmed that at Gaviotas, dreams that hang on long enough could finally come true.

"I'll be damned. You actually did it."

We were entering a tall parabolic tent, sort of a canvas Quonset hut appended to the Gaviotas resin factory. Inside, tethered by nylon ropes was the biggest silver bullet I'd ever seen. At last, the Gaviotas dirigible.

"You like it? We built it ourselves, with no technical assistance."

The zeppelin, sleek and gorgeous, was 65 feet long and 10 feet wide. It was made of mylar, polyurethane, and polyethylene, purchased with $50,000 in grants from the United Nations and the Colombian government. It was not, however, serving the purpose originally contemplated: lighter-than-air shipment of pine resin across the llanos, to save gasoline.

"We hadn't grasped the technical difficulty, not the cost, of a dirigible big enough to carry that much freight, plus a crew." This one, operated by remote control, carried only an infrared video camera that monitored hot spots in the forest before they turned into forest fires. Aloft, it could scan 4,000 hectares at once.

The idea of deriving hydrogen from water to inflate it had also been scotched when the government meteorological station that once launched weather balloons at Gaviotas relocated, taking its electrolyzer with it. "We're using helium. Maybe some day we'll figure out how to fill it with hot air by tapping factory exhaust."

We had already toured the co-generating boiler, fired by culls from the Gaviotas forest, that provided heat to process pine resin and whose vaporous exhaust spun a turbine that electrified the entire village. In the processing plant, I found Hernán Landaeta still in charge, directing a dozen men hauling pallets heaped with aromatic bags of newly harvested pine resin. He guided me over to a new Gaviotas brainchild.

It was a tubular steel column, which, Landaeta explained, functioned much like an oil refinery distillation tower. But instead of fractionating crude petroleum into different densities ranging from tar to gasoline to natural gas, it separated refined pine resin into eighteen new potential products, from natural chewing-gum base to varnish-grade colophony to pine-oil disinfectant.

These new derivatives were born of necessity. Until 2007, the Colombian paint industry had purchased nearly all of Gaviotas' annual resin harvest. Then China entered the market in typically colossal fashion: with 500,000 tons of resin from pine forests in its rural west, lately being subsidized by Beijing to develop at the breathtaking pace of its eastern provinces. Though less refined than high-grade Gaviotas colophony, the Chinese product was adequate for paint manufacture, and it sent prices plummeting.

"So we've had to diversify," Paolo said, cradling a flask of pine-oil floor cleaner in his thick fingers. "We're think of manufacturing our own brand of sugar-free gum. How does *Chicle Gaviotero* sound?

After what I saw next, it sounded possible. The former Gaviotas hospital had been transformed—though not, as they'd hoped a decade earlier, into a medicinal plants research center: The violence whipping through the Colombian llanos had deferred that project as well. But another idea discussed back then now promised to join pine resin, solar panels, windmills, and pumps as a viable way to sustain their village.

The ex-hospital had become a bottling plant for Gaviotas mineral water—whose purity, according to a sheaf of tests by an eminent Tokyo laboratory, was exceptional. The plant's young director, Andrea Beltrán, who led me through bottle purification chambers that were once maternity rooms, I remembered as Abraham Beltrán's shy pre-teen daughter. She had been to the city for school, but returned to Gaviotas as soon as she graduated.

The *maloca*, where Guahibo Indians once convalesced in hospital hammocks under a thatched roof, was missing. "We had to minimize possible sources of organic contaminants," Andrea explained.

No one had ever gotten sick from Gaviotas water, but this was a market necessity: Gaviotas Agua Natural Tropical had to follow federal sanitary production rules, because much of it was destined for restaurants. In the former surgery, a team of women glued labels onto bottles that read "Wok"—a fashionable Begotá chain that served Asian food. More sophisticated bottling and labeling equipment was on its way. They'd just signed a contract, Lugari said, with Colombia's most famous brand: Café de Colombia's trademark coffee grower, Juan Valdez.

A national chain by that name, with outlets reminiscent of Seattle-based Starbucks but selling only Colombia's finest, would now also offer Agua Natural Tropical from Gaviotas. Paolo handed me a bottle with a logo showing the mustachioed *cafetero* with his sombrero and burro.

"Water from forty meters deep, filtered through a hundred kilometers of sand, with no agricultural chemicals anywhere near and surrounded by a forest," boasted Lugari, draining his in two swallows. "Have you ever tasted better?"

The bottle itself was yet another innovation: instead of the typical cylinder, it was a four-sided prism with two round indentations on one side matched by rounded protrusions on the opposite face. The design allowed bottles to interlock—an inspiration that came to Paolo one mornings as he watched his housekeeper's son playing with Lego blocks. Besides being far easier to ship and stack on store shelves, instead of being thrown away, children were collecting them.

"They call them *Legos de los pobres*—poor people's Legos," Lugari told me. At Gaviotas they were also filling them with sand to use as bricks to build walls.

Ingenious, I admitted. But they were still plastic.

"We use recyclable polyethylene."

"But how many people actually recycle them?"

"It's part of our contract with the Juan Valdez restaurants. Every time we deliver a shipment, we take back all the empties that people don't save for their kids."

A smart plan, and the trick of turning potential trash into toys was surely charming. But plastic was still a petroleum product, one that nature had yet to learn how to digest, and sooner or later most, if not all, of these bottles would end up in waste streams that had nowhere good to flow.

"True," said Paolo. "Unless. …"

I'd seen the flash of an idea illumine his eyes before. "Unless what?

"Unless we learn to make biodegradable plastic out of renewable palm oil."

"That's the plan?"

"The plan is to try. Look," he said, "if you drank two bottles of this water a day for twenty-five years, we calculate that the money you spend would help us regenerate nine hectares of forest that absorb 165 tons of carbon dioxide. Even if we can't make them from palm oil, their impact will clearly be compensated a thousand times by all the trees we're planting."

Possible. It was at least clear that selling gourment water could be a financial windfall to Gaviotas as it tried to survive and even thrive—sustainably, no less—on an increasingly precarious global stage. Happily, we were interrupted at this point by the appearance of Teresa Valencia, *la profesora,* who'd come to fetch us for lunch.

Lush gallery jungle shaded the road that led to the center of Gaviotas. Green parrots and saffron kiskadees zipped through air thick with humidity, signaling that the rains were nearing. Next to the small bridge that crossed the Caño Urimica, two Gaviotans were bathing their baby son in the shallow stream: nearby stood a pair of whistling herons, their long necks swiveling like periscopes as we passed. Somewhere high in the canopy, I could hear monkeys leaping.

Alonso Gutiérrez, Teresa told me, was still employed in the coffee industry, returning to Gaviotas as work permitted. Sometimes they met in Villavicencio, where their daughter, Natalia, was now in secondary school.

"She was just a baby—"

"I know. I wish you could see her. She's a real *gaviotera.*"

"Like her mom."

A few years earlier, when Juan David Bernal's recurring health problems forced his family to return to Bogotá, Teresa seemed the logical choice to replace Gonzalo as coordinator—except, as she pointed out, there really wasn't need for one. Everyone knew what needed to be done, and everyone did his or her job. Still, many village administrative details had fallen to her, and a new teacher had arrived from Cartagena to run the primary school. Teresa still taught: at lunch, a dozen schoolchildren flocked around her to report on the morning's art projects.

We ate fish salad made from native *cachama* that Gaviotas now raised in its own ponds. I was surprised that the vegetables weren't hydroponic: During the worst of the recent civil disorder, with the whole country's finances reeling, they'd decided to forgo the cost of hydroponic nutrients. Instead, they bought lettuce, carrots, tomatoes, and leeks from neighbors who'd learned to coax them from savanna soils fortified with kitchen ashes and manure from chicken and hogs.

It was hard to imagine Gaviotas without hydroponics, and apparently I wasn't alone. Everyone missed the ready supply of spinach, radishes, cilantro, parsley, onions, and beets, which didn't do well in local truck gardens. "We'll bring it back, as our new products start earning," said Paolo. "We've been researching organic hydroponics that use no artificial chemicals. If we can make soil in our forest, why can't we make our own nutrients? It's another technology we can share with our neighbors."

I'd learned to take any idea broached at Gaviotas seriously, no matter how improbable. Even those that failed often led to something that worked. On the lawn by the community center, Pompilio showed me the latest, in response to a neighbor's request for a really cheap pump to tap shallow groundwater for his garden. They'd sunk an eight-inch PVC pipe four meters into the ground, and attached a lever that lifted and dropped a flap covering the pipe's exposed end. Each time Pompilio pumped, water gurgled closer until finally it gushed over the top. It was little more than a giant soda straw, and worked on basically the same principle.

The population of Gaviotas, around 200, was nearly the same as when I'd last visited, but its economy sustained more than 2,000 people in the surrounding area, many of them indigenous Guahibos. Llaneros still brought children to the Gaviotas school. Some had grown and stayed, others went to the city and remained Gaviotans, sometimes working in the Bogotá factory that produced windmills, pumps, and solar collectors, and now biodiesel. Although a subsequent Colombian president had removed the Gaviotas solar water heaters that ex-president Belisario Betancur installed years earlier at his official residence—reminiscent of Ronald Reagan yanking Jimmy Carter's solar panels from the White House roof—Lugari considered it a hopeful portent that among their most recent solar energy customers in Bogotá was the U.S. Embassy.

Obligatory solar water heaters, networks of bike lanes, redoubled mass transit, rooftop agriculture, vegetable-based plastics, urban services within walking distance, and trees planted wherever possible: He hadn't given up believing that existing cities could become sustainable. But the chance to start from scratch in this big empty savanna was clearly Paolo Lugari's passion.

"I still dream of us building Odisea," he said as we headed to the airstrip, escorted by a throng of bicycle-mounted Gaviotans. "People were scared to move way out here. Or they went to Arauca and Casanare because they thought they could make money in the oil fields. Or they just settled on the first vacant land they found—there's still so much before you reach Gaviotas. But it will happen. Today we started growing energy in our pine resin forest. If we can do that, we can grow food, too. Some rice varieties don't need irrigation. Same with cassava, corn, bananas. Even soy. We'll fertilize with residues from the biodiesel plant. And we'll nurture mycorrhizas—they can make everything grow here."

Our plane made a circle over the spreading Gaviotas forest. The pines around the verdant *morichal* at Odisea were no longer the young shrubs I'd last seen there, but tall trees. Between them, the surrounding dry yellow llano grass disappeared beneath a deep green tangle of native flora. We made a last pass over the village, dipping a wing to the waving cyclists below, then banked toward the Andes. The savanna before us seemed big as an ocean—except oceans don't burn, and this one was ablaze, dusting the sky with chalky haze.

Lugari turned to me from the copilot's seat. "Every year they burn the llano before the rains, to free potassium so that poor savanna grass can support a few scrawny cattle. Thousands of tons of CO_2 released just to put one cow on every hundred hectares."

Through the smoke, I could see a jagged orange line creeping to the edge of the big *caño* at Carimagua. I pointed to some *moriche* and *ceje* palms in flames.

"So stupid." Paolo gestured at the charred ground below." Have you read that article I gave you?"

I had. The Colombian business magazine *Dinero* estimated that with a million hectares of African palm, Colombia could meet all its demand for diesel.

"There's at least six million hectares burning down there. There are forty million more in Colombia, empty and available for growing palms, pines, and all kinds of food. If people want animals, we can use low-impact African sheep that don't graze down to the roots like goats. If we reforest all the tropical savannas, we can soak up CO_2—and, by making fuel from palm oil, we can stop adding more."

"I thought using palm oil for biofuel is ridiculously expensive."

"At today's price, yes. But growing it in a polyculture like ours, on land that's otherwise wasted, isn't ridiculous: it's intelligent. It's a crime to tear down Africa's jungles to grow fuel. In Europe or the United States, it's absurd to sacrifice food for biodiesel from corn, sugar beets, or sunflowers. But Latin America's savannas are the most unexploited soils in the world. We could plant all year and continually harvest different crops from a healthy, biodiverse system."

The smoke thickened around us. We were nearing the Andes, but haze had swallowed the mountains and the civilization climbing their skirts. An hour earlier, I'd stood in a clean, fragrant forest that felt so vast—millions of trees not even thirty years old, yet already so high and so varied. But now Gaviotas seemed miniscule compared to the world of troubles that surrounded it. Was it truly an example to show us how to reinvent our world? Or was it just a sweet, irrelevant anomaly—an island of sanity because of its isolation, like a sage whose safeguards his wisdom in a remote cave?

The pilot taxied to a stop, then cut the engine. The predominant fragrance was once again tarmac and fumes.

Paolo sighed. "You know," he said, "I've never wanted Gaviotas to be some kind of eco–doll house, or pilot project, or some toy for NGOs. I want it to show the world how to fortify an ecosystem. Sometimes I think biodiesel may be the most important chance we have. People would be planting, not exhuming energy. They'd be restoring the planet's living skin. The atmosphere's equilibrium depends on the planet's biomass. Biofuel may be the only way to keep atmospheric chemistry in balance and global warming in check."

Was he dreaming? Could we really have our engines and our world, too?

"We have to keep dreaming," Paolo answered. "If you're not dreaming, then you're asleep. The real crisis isn't a lack of resources: It's lack of imagination."

The flash was back in his eyes. "Just imagine," he said, a smile starting to tug at his gray beard. "Imagine if everyone on earth were required to plant at least three trees. …"

Alan Weisman
May, 2008

For ongoing information:
http://www.friendsofgaviotas.org
http://www.centrolasgaviotas.org